4th International Conference on Reliability Engineering (ICRE 2019)

IOP Conference Series: Materials Science and Engineering
Volume 836

Rome, Italy
20 - 22 November 2019

ISBN: 978-1-7138-2679-8
ISSN: 1757-8981

Printed from e-media with permission by:

Curran Associates, Inc.
57 Morehouse Lane
Red Hook, NY 12571

Some format issues inherent in the e-media version may also appear in this print version.

Institute of Physics
Dirac House, Temple Back
Bristol BS1 6BE UK

Phone: 44 1 17 929 7481
Fax: 44 1 17 920 0979

techtracking@iop.org

Additional copies of this publication are available from:

Curran Associates, Inc.
57 Morehouse Lane
Red Hook, NY 12571 USA
Phone: 845-758-0400
Fax: 845-758-2633
Email: curran@proceedings.com
Web: www.proceedings.com

TABLE OF CONTENTS

IOP Conf. Series: Materials Science and Engineering **836** (2020) 011001 doi:10.1088/1757-899X/836/1/011001

PREFACE

On behalf of the organizing committee and technical program committee, we would like to extend to you a warm welcome to 2019 4th International Conference on Reliability Engineering (ICRE 2019) in Rome, Italy.

The aim objective of ICRE 2019 is to provide a platform for researchers, engineers, academicians as well as industrial professionals from all over the world to present their research results and development activities in Reliability Engineering. This conference provides opportunities for the delegates to exchange new ideas and application experiences face to face, to establish business or research relations and to find global partners for future collaboration.

The proceedings present a selection of 10 papers submitted to the conference from universities, research institutes and industries. We continue to build on the theme of reliability engineering, with keynotes and papers on this important and growing topic. Every year at ICRE we strive to organize a conference that is technical strong while offering a great attendee experience. We look forward to your survey feedback in continuing to keep this conference a great experience in future years. Welcome to ICRE and we hope you will enjoy both the conference and the sunshine in Rome.

Organizing Committee of ICRE 2019

IOP Conf. Series: Materials Science and Engineering **836** (2020) 011002 doi:10.1088/1757-899X/836/1/011002

Conference Committee

Steering Committee

Prof. Francesco Di Maio, Politecnico di Milano, Italy

Prof. TANG Loon Ching, National University of Singapore, Singapore

Prof. Yi Ding, Zhejiang University, China

Prof. Rui Kang, Beihang University, China

Conference General Chairs

Prof. Michael Pecht, Center for Advanced Life Cycle Engineering at the University of Maryland, USA

Prof. Enrico Zio, Politecnico di Milano, Italy

Program Chairs

Prof. Liudong Xing, University of Massachusetts, USA

Prof. Sebastian MARTORELL, Universitat Politècnica de València, Spain

Prof. Jin Wang, Liverpool John Moores University, UK

Prof. Dr. Francesca Saglietti, University of Erlangen-Nuremberg, Germany

Technical Committee

Prof. Antoine Rauzy, NTNU, Norway

Assoc. Prof. Yanhui Lin, Beihang University, China

Assoc. Prof. Xiaoyang Li, Beihang University, China

Assoc. Prof. Sompoap Talabgaew, King Mongkut's University of Technology North Bangkok, Thailand

Assoc. Prof. Shinji Inoue, Kansai University, Japan

Assoc. Prof. Rui Peng, University of Science & Technology Beijing, China

Assoc. Prof. Rong Pan, Arizona State University, USA

IOP Conf. Series: Materials Science and Engineering **836** (2020) 011002 doi:10.1088/1757-899X/836/1/011002

FirstName	LastName	Institution	EmailAddress
Andres Raul Herrera	Iriarte	Universidad Pontificia Bolviariana, Colombia	andres.herrerai@upb.edu.co
Stephanie	Chahine	LARIS Laboratory, Angers University, France	stephanie.chahine@univ-angers.fr
Francesco	GHERARDINI	University of Modena and Reggio Emilia, Italy	francesco.gherardini@unimore.it
Alexandra	Khalyasmaa	Ural Federal University, Russia	lkhalyasmaa@mail.ru
Alexandra	Khalyasmaa	Ural Federal University, Russia	lkhalyasmaa@mail.ru
Stanislav A.	Eroshenko	Ural Federal University, Russia	stas_ersh@mail.ru
Olga	Shabunina	Ural Federal University, Russia	shabunina-olga@yandex.ru
Daria	Shatunova	Ural Federal University, Russia	d.v.shatunova@urfu.ru
Titiya	Lapwattanamongkol	Chulalongkorn University, Thailand	pam.titiya@gmail.com
Yichao	He	Nanjing University of Science and Technology, China	yichaoh@163.com

IOP Conf. Series: Materials Science and Engineering **836** (2020) 011004 doi:10.1088/1757-899X/836/1/011004

Peer review statement

All papers published in this volume of *IOP Conference Series: Materials Science and Engineering* have been peer reviewed through processes administered by the proceedings Editors. Reviews were conducted by expert referees to the professional and scientific standards expected of a proceedings journal published by IOP Publishing.

IOP Conf. Series: Materials Science and Engineering **836** (2020) 012001 doi:10.1088/1757-899X/836/1/012001

Study of sensitivity for strain-based structural health monitoring

A Herrera-Iriarte[1], J Alvarez-Montoya[1] and J Sierra-Pérez[1]

[1]Grupo de Investigación en Ingeniería Aeroespacial, Universidad Pontificia Bolivariana, Medellín, Circular 1 # 70-01, Colombia

andres.herrerai@upb.edu.co

Abstract. One of the available methodologies for structural health monitoring (SHM) is based on strain field pattern recognition where, through the use of sensors capable of measuring strain on discrete points and machine learning techniques, it is possible to detect a damage event. In this study, strain data from fiber optic sensors (FOS), in particular fiber Bragg gratings (FBG), acquired through two experiments are used: an aluminum beam with 32 FBGs and CFRP beam provided with 20 FBGs, which serves as the main wing's structure of an unmanned aerial vehicle (UAV). Both structures were subjected to dynamic loading for a pristine condition and later, for artificially damaged conditions. In the experiments presented in this paper the beams were provided with different amounts of sensors which were removed one by one in order to analyze the sensitivity of the damage detection methodology based on PCA to a change in the number of sensors. The results demonstrated that there are few sensors that contribute mostly to the methodology's performance, these sensors are validated to be the ones located near the analyzed damage condition. Therefore, this study is the first step into the development of methodologies of damage localization using strains.

1. Introduction

Monitoring techniques are important on the reliability of several industries such as aerospace, civil construction, nuclear and automotive. Since any real system or structure wear out, it is paramount to identify how and when a catastrophic failure will occur, guaranteeing the serviceability of the system over time and, therefore, avoiding human and economic losses. Having as example an airline, structural maintenance represents 33% of the total maintenance cost, at the same time maintenance is around 20% of total costs of an airline, therefore, it is important to find a solution that reduces costs while not affecting reliability [1].

Structural health monitoring (SHM) refers to the implementation of a damage detection strategy through a network of sensors in order to monitor and extract damage-sensitive measurements to determine the integrity of a structure or a system over time [2]. This approach allows the shift from a schedule-driven maintenance regime to a condition-based maintenance (CBM) [3].

That is why, nowadays, it has raised the interest among industrials and academics in SHM due to the most quantity of benefits that this could bring to traditional monitoring techniques such as non-destructive testing (NDT). These benefits include real-time damage detection that could lead to a reduction of safety factors for composite materials (in order to design more efficient and lighter structures), damage severity and location assessment and maintenance cost and time reduction for aerospace applications by replacing traditional NDT which consumes a lot of resources and time.

Among the available methods for performing SHM, one of the most promising ones is strain-based SHM using fiber optic sensors (FOS). These strain sensors have the capability of being embedded into the material in the manufacturing process without being invasive and degrade the performance of the system. Once sensor information is obtained, data processing has to be carried out in order to decision-making, for this, some of the most successful methods are the application of pattern recognition and machine learning algorithms, and one of the most used machine learning techniques for this purposes is principal component analysis (PCA) [4].
PCA consists of a statistical method in which a data with several dimensions can be reduced to a desirable number of components that has the characteristic of showing the highest possible variability with the established components. SHM systems usually require a large number of sensors acquiring data over a large period, therefore, it is necessary to use dimensionality reduction techniques in order to ease the processing [2].
The authors have been working on the use of PCA to perform damage detection from discrete strain measurements acquired by fiber Bragg gratings (FBGs), a type of FOS. Such approaches have demonstrated suitable results in several experiments such as an aluminum beam subjected to dynamic loading [4] and an unmanned aerial vehicle (UAV) composite beam under real operational conditions [5]. Although several experiments have been performed by using strain data in conjunction with PCA, there are no studies that quantify changes on damage indexes as a function of the number of used sensors. One important question when designing such SHM systems, is the quantity of sensors required in order to have an appropriate sensitivity. In this study, data from two experiments (aluminum beam and UAV composite beam) were used in order to perform a sensitivity analysis with a variable number of sensors. The analysis is based on removing specific sensor data and analyzing whether left sensors can detect damages. The aim is to provide tools that allow designers to decide the adequate number of sensors in a desired application in order to design SHM systems with higher cost-benefit ratios. In addition, this study can provide useful information for damage localization since near-damage sensors are supposed to contribute more to the damage detection capability [6].

2. Background

There are many ways of performing SHM aside strain-based analysis, a few are discussed below. One popular method referred to vibration-based SHM is acoustic emissions in which sensors are special "microphones" that "listen" the structure and predict with this noise if a failure is about to occur. This method has the disadvantage of having a lot of noise on their measurements and required more robust equipment compared with strain-based analysis using FOS. Another used method is acceleration-based SHM in which a lot of accelerometers are installed around a structure, this method is an alternative to measure strains indirectly. Temperature is another possible variable to measure in order to detect damages, but this method does not have enough sensitivity for detecting small damages and it is mostly used for detecting humidity into materials or big scale damages [7].
PCA as a machine learning method is used not only in SHM applications, it has a lot of additional applications. It is used, in general, for image identification (facial recognition and image classification as well), fingerprint scanning, recently for sound recognition [8] and the studies in medical sciences for heart rate monitoring [9]. As mentioned before, PCA tents to reduce components of a multidimensional array with the most possible quantity of information in the components obtained.
It is important to identify the final dimension of the matrix when PCA is applied, for example, if initially there is a matrix of 4000 x 20, it means that there are 4000 measures (or samples) and 20 sensors (or features), with a PCA algorithm it is intended to obtain a 4000 x 3 matrix, it means that the initial 4000 x 20 matrix is projected from a 20-dimensional space to a 3-dimensional space. For this, the following equations are used:

$$coeff = \lambda(cov(A)), \tag{1}$$

$$score = A \times coeff, \tag{2}$$

where, A is the initial matrix and $score$ the resultant matrix after PCA is applied. In equation (1) projection matrix $coeff$ are the eigenvectors of covariance [8].
For all experiments the data is divided into a baseline matrix (a healthy condition matrix) and a D0 matrix (that is a undamaged data matrix used for validation). Finally, D1, D2, D3 ... DN, are the data matrices for different damage conditions. Once obtained the PCA reduction matrix, it proceeds to classify as damage or no damage. This is done with help of Q index which is calculated using following equations:

$$\bar{X} = score \times coeff^{T}, \tag{3}$$

$$E = A - \bar{X}, \tag{4}$$

$$Q = \sqrt{E \times E^{T}}, \tag{5}$$

Where Q index is a vector equal to measures number [4].
In machine learning techniques; it is common to perform receiving operating characteristic (ROC) analysis in order to determine the performance of a classifier. Within this analysis it is important to define several metrics such as true positive (TP), data which belong to a detected damage that is a real damage, true negative (TN), when there is no damage detected and the structure is undamaged, false positive (FP), when there is no a damage and algorithm detect damage and false negative (FN) when a damage exists and the algorithm does not detect it.
$F1$ score values determine the accuracy of a classifier, it takes a value between 0 and 1 being 0 the worst and 1 the better case. It is commonly used due to it allows performance comparison combining the above discussed metrics in one.

$$F1 = \frac{2 \times TP}{2 \times TP + FP + FN} \tag{6}$$

3. Experimental setup

As mentioned before, two experimental data sets were used, one from an aluminum beam submitted to cyclic bending loads and the other from the SMARP UAV structure made of CFRP.
The aluminum beam structure experiment was performed under a controlled environment obtaining a true positive true rate of 100% and a false positive rate of 1.28% [4].
The beam had a rectangular cross section of 20 mm x 40 mm and a total length of 1200 mm with 32 FBG sensors bonded to its surface. Cyclic bending loads were applied into a cantilevered beam by attaching it to an AC motor provided with a connecting rod. A total of six artificial damages were induced by drilling holes which produced a small local change in stiffness. In figure 1, a depiction of the damage locations is presented, where the damage selected preliminary for the study is highlighted in red (D5). For a detailed description of the experiment see [4].
The SMARP UAV beam experiment was performed using 20 FBG sensors embedded in the SMARP UAV structure. Real operation tests (flights) were carried out obtaining a true positive rate of 99.7% and a false negative rate of 0.3% [5].
The UAV flight data was sent in real time to a ground station using 2.4GHz WLAN wireless system, during the tests a lot maneuvers were performed, varying roll, pitch and yaw angles, speed, vertical speed. The simulated damages in the structure were performed by bonding plates of a stiffer material on certain points of the structure. In figure 2, a depiction of the damage selected for the study is detailed in conjunction with the near sensor locations. For a detailed description of the experiment see [5].

4. Results and discussion

Results of Q index for both experiments are presented in figures 3 and 4.

IOP Conf. Series: Materials Science and Engineering **836** (2020) 012001 doi:10.1088/1757-899X/836/1/012001

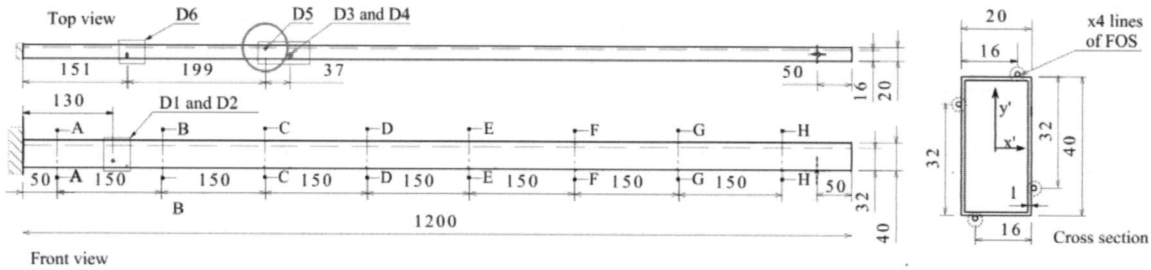

Figure 1. Damage localization for aluminum beam [4].

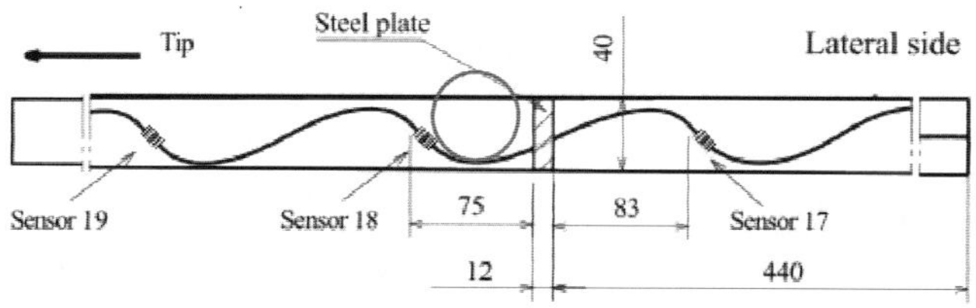

Figure 2. Damage localization for CFRP beam [5].

As can be seen in figure 3a and figure 4a damage cases are over the thresholds, it means that damages are detected, however, for figure 3b and figure 4b damages are under thresholds so there are no damages detected. It can be appreciated that by reducing the number of sensors the capability of the methodology of detecting damages is seriously affected. Based on such figures, the $F1$ score is calculated for each experiment and for different number of sensors (see figures 5a and 5b). It is important to highlight that sensors were removed from nearer to farther sensor from damage. It can be seen that in both cases there is a number of sensors from which there is not considerable changes in the $F1$ score, which means that said sensors are not contributing considerably in the damage detection process.

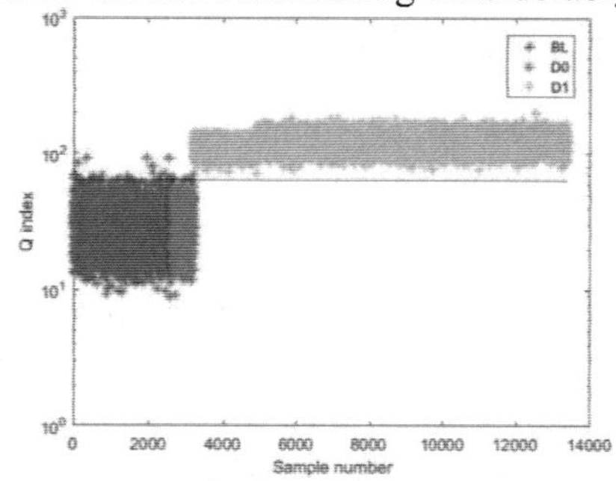

Figure 3a. Q index for 20 operating sensors, CFRP experiment.

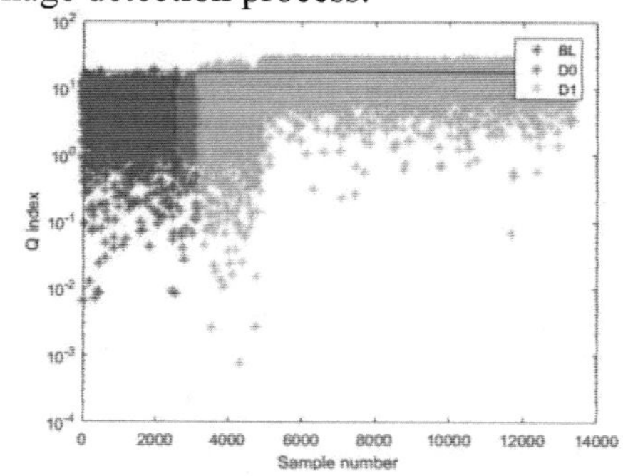

Figure 3b. Q index for 3 operating sensors, CFRP experiment.

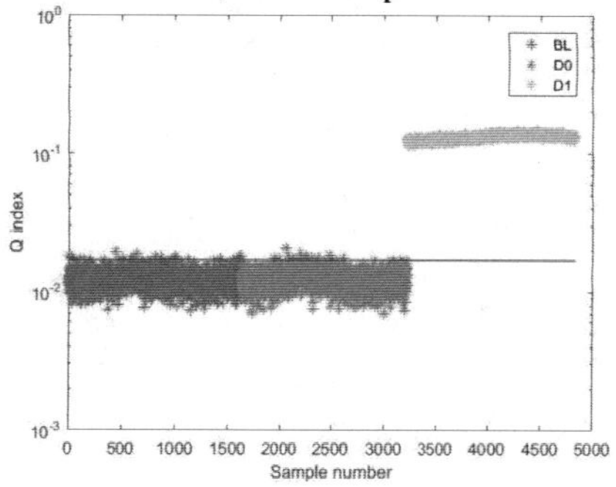

Figure 4a. Q index for 31 operating sensors, aluminum beam experiment.

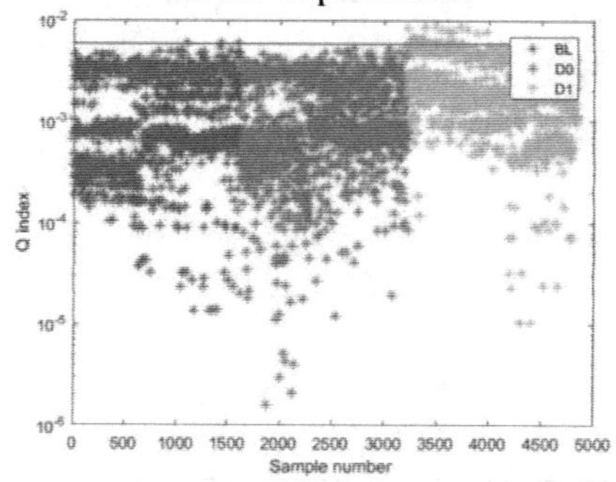

Figure 4b. Q index for 4 operating sensors, aluminum experiment.

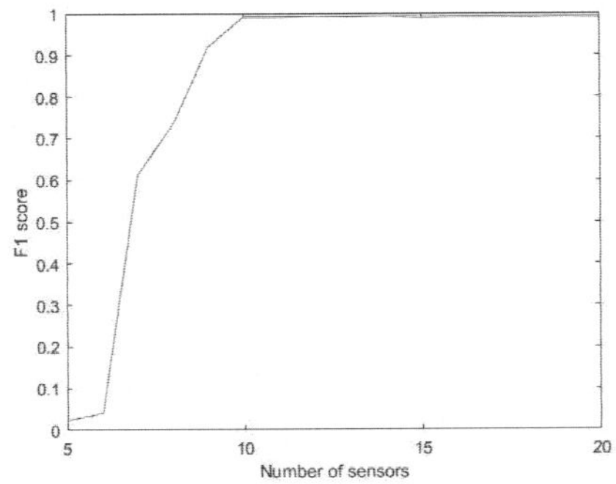

Figure 5a. $F1$ score vs sensors number for aluminum beam.

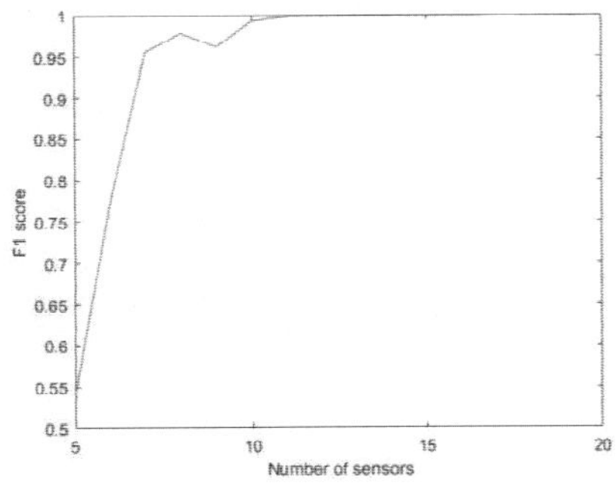

Figure 5b. $F1$ score vs number of sensors for CFRP beam.

5. Conclusions

For $F1$ scores plot it can be observed that when sensors number increases there is a point in which $F1$ score does not increase significantly, this point indicates an ideal number of sensors that could be installed in order to have a balance in cost and sensitivity. For the CFRP beam case when ten sensors are operative there is a 0.99 rate for $F1$ score what indicates an acceptable reliability and for the case of the aluminum beam, the score after ten sensors removed is 0.99. For both cases, the $F1$ score does not increases significantly when there are more than ten sensors, it means that, for these two damage examples and experiments, a damage can be detected for multiple sensors especially those that are located near the damage, this having into account that sensors were removed from nearer to farther from the damage location. This represents valuable information toward damage localization using strain data and machine learning techniques since the location of the damage can be delimited to the zone where the sensors with the most considerable contributions to the $F1$ score are presented the contribution of the sensors to the performance of the methodology.

6. References

[1] IATA 2017 *AIRLINE MAINTENANCE COST*
[2] Farrar C R and Worden K 2010 An introduction to structural health monitoring *CISM Int. Cent. Mech. Sci. Courses Lect.* **520** 1–17
[3] Fu H, Dimitrios B, Sharif Khodaei Z and Ferri Aliabadi M H 2018 Structural Health Monitoring Unit for Condition Based Maintenance of Composite Structures *International Symposium on Structural Health Monitoring and Nondestructive Testing*
[4] Sierra-Pérez J, Torres-Arredondo M-A A M-A and Alvarez-Montoya J 2018 Damage detection methodology under variable load conditions based on strain field pattern recognition using FBGs, nonlinear principal component analysis, and clustering techniques *Smart Mater. Struct.* **27** 015002
[5] Alvarez-Montoya J, Torres-Arredondo M-. and Sierra-Pérez J 2018 Gaussian process modeling for damage detection in composite aerospace structures by using discrete strain measurements *Proceedings of the 7th Asia-Pacific Workshop on Structural Health Monitoring, APWSHM 2018* pp 710–8
[6] Ruiz M, Mujica L E, Sierra J, Pozo F and Rodellar J 2018 Multiway principal component analysis contributions for structural damage localization *Struct. Heal. Monit.* **17** 1151–65
[7] Holnicki-Szulc J 2008 *Smart technologies for safety engineering* (J. Wiley)
[8] Aithal M B, Gaikwad P R and Sahare S L 2015 Speech Enhancement Using PCA for Speech and Emotion Recognition **4** 6–12
[9] Qureshi N A, Suthar V, Magsi H, Sheikh M J, Pathan M and Qureshi B 2017 Application of Principal Component Analysis (PCA) to Medical Data *Indian J. Sci. Technol.* **10** 1–9

IOP Conf. Series: Materials Science and Engineering **836** (2020) 012002 doi:10.1088/1757-899X/836/1/012002

Probabilistic analysis of fatigue crack growth using efficient surrogate model

S Chahine[1], **H Riahi**[1] **and D Bigaud**[1]

[1]LARIS Laboratory, Angers University, 62 Avenue Notre Dame du Lac, F-49000 Angers, France

Stephanie.chahine@univ-angers.fr

Abstract. In this paper probabilistic analysis of a fatigue crack growth process governed by the Walker law is performed. Sensitivity of its parameters to uncertainties related to the material properties is considered. Based on statistical analysis of experimental data, they are modelled as independent normal random variables. The effect of these uncertainties on the fatigue crack growth behaviour of Centre Cracked Plate specimen is assessed using Monte-Carlo simulations and surrogate model based on polynomial chaos expansion. An efficient truncation scheme allows to discard the high order interactions having a weak effect on the mechanical response and consequently reduce the number of finite elements calls when identifying the unknown coefficients of the polynomial chaos expansion. The statistical moments namely the mean and the standard deviation in addition to the probability density function of the mechanical response are derived with a good accuracy. The obtained results show that the fatigue crack growth life is significantly affected by the uncertainties on the material properties since its coefficient of variation reaches 21% and that it follows a log-normal distribution law.

1. Introduction

For many years now, researchers [1] have pointed out the stochastic character of the fatigue crack growth. The variability of the fatigue crack growth life was attributed to the uncertainties observed on the material properties, the geometrical parameters and the loading conditions. It is clear that to obtain a safe design, these sources of uncertainties should be taken into account. The probabilistic approaches could address this issue, but unfortunately they suffer until now from some limitations to solve real-life engineering problems. Indeed, some of them are purely statistic [2] and subject of criticisms since they are not able to describe the physical phenomena related to the fatigue crack growth. Other approaches, which are mainly based on the *probabilization* of the fatigue crack growth rate [3], are time consuming, because their efficiency is affected when the stochastic dimension (i.e. the number of random variables representing the uncertain parameters) of the problem is high, and/or the mechanical model representing the fatigue crack growth is also time consuming. Recently, a perturbation series expansion based-method is developed [4, 5], which is only accurate when the crack growth process can be described through simple deterministic mathematical models. Long et *al* [6] have developed an innovative approach which allows, on the one hand to reduce the number of mechanical model calls by the use of efficient integration scheme and on the other hand to take into account the effect of statistical dependence [7] between crack growth uncertain parameters. As an alternative, this paper presents an efficient uncertainty propagation method based on Polynomial Chaos Expansion (PCE) to deal with fatigue crack growth problem. The aim of this method is to reduce the number of simulations required to quantify uncertainties.

IOP Conf. Series: Materials Science and Engineering **836** (2020) 012002 doi:10.1088/1757-899X/836/1/012002

2. Uncertainty propagation based on PCE

The uncertainty propagation aims to characterize, in a statistical manner, the variability observed on the behaviour of a physical model when some of its input parameters are random quantities. Let us consider a mechanical model having n uncertain input parameters gathered in the n-dimensional vector $\boldsymbol{x} = \{x_1, x_2, \dots, x_n\}^T$, and let $\boldsymbol{X} = \{X_1, X_2, \dots, X_n\}^T$ the probabilistic model of $\boldsymbol{x}$ which is fully described by its joint probability density f_X. Mathematically speaking, the response y (i.e. stress, crack length, fatigue life) of the mechanical model, which is, without any loss of generality, considered here as a scalar, can be linked to the input parameters $\boldsymbol{x}$, through the mapping $y = \mathcal{M}(\boldsymbol{x})$. Since $\boldsymbol{x}$ is random, consequently the mechanical response y is also a random quantity, with associated probabilistic model Y which can simply be a random variable. According to the PCE method [8], the pth order approximation of the mechanical response y, written in the standard random space where the n-dimensional random variable $\boldsymbol{X} = \{X_1, X_2, \dots, X_n\}^T$ is substituted by an n-dimensional standard normal variable $\boldsymbol{U} = \{U_1, U_2, \dots, U_n\}^T$ using a suitable iso-probabilistic transformation $\boldsymbol{X} = T(\boldsymbol{U})$, reads:

$$y \approx \widehat{M}(\boldsymbol{X}) = \widehat{M} \circ T(\boldsymbol{U}) = \sum_{|\alpha| \le p} a_{\boldsymbol{\alpha}} \Psi_{\boldsymbol{\alpha}}(\boldsymbol{U}) \tag{1}$$

where $\boldsymbol{\alpha} = \{\alpha_1, \alpha_2, \dots, \alpha_n\}$, with $|\boldsymbol{\alpha}| = \alpha_1 + \dots + \alpha_n$, are multi-index, $a_{\boldsymbol{\alpha}}$'s are unknown deterministic coefficients and $\Psi_{\boldsymbol{\alpha}}$'s are multivariate Hermite polynomials which form an orthogonal basis with respect to the normal probability density and can be obtained by tensorizing n univariate Hermite polynomials $\{\Psi_{\alpha_k}\}_{k=1}^{n}$.

The unknown coefficients $a_{\boldsymbol{\alpha}}$ can be obtained by performing regression analysis based on numerical plan of experiments. The estimates of the $a_{\boldsymbol{\alpha}}$'s are computed through the minimization of the mean square error of the polynomial approximation $\widehat{M}$. The regression problem reads:

$$\text{Find } \widehat{\boldsymbol{a}} \text{ that minimizes } \mathbb{E}\left[\left(a^T \Psi_{\boldsymbol{\alpha}}(\boldsymbol{U}) - M(\boldsymbol{U})\right)^2\right] \tag{2}$$

where $\mathbb{E}[.]$ denotes the mathematical expectation.

Let $\mathcal{T}_n^p = \{\boldsymbol{\alpha} \in \mathbb{N}^n \; such\ that \; |\boldsymbol{\alpha}| \le p\}$ be a set of index associated with a PCE of degree p. Hence, the number of terms in the PCE is $card(\mathcal{T}_n^p) = \binom{n+p}{p} = \frac{(n+p)!}{n!p!}$. The convergence rate of the estimates of the unknown coefficients can be enhanced by the use of suitable plan of experiments such those based on low discrepancy random sequences, which ensure a better representation of the random space. Here, Gauss-Hermite integration points are used to construct the plan of experiments. Firstly, $p + 1$ one-dimensional Gauss-Hermite integration points are generated. Then n-tuples representing all possible combinations from the $p + 1$ one-dimensional Gauss-Hermite integration points are computed. Since there are too many combinations, only the first $2 \times (card(\mathcal{T}_n^p) + 1)$ points are selected to construct the plan of experiments, after sorting the n-tuples in ascending order according to the distance to the origin. Figure 1 compares the obtained 2D plan of experiments for PCE order $p \in \{2,4,6\}$.

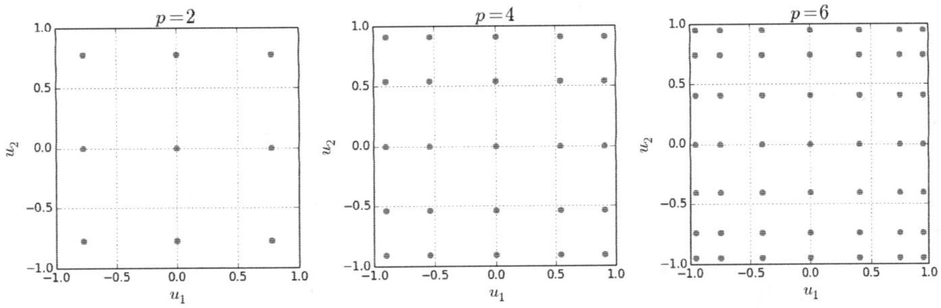

Figure 1. 2D plan of experiments based on Gauss-Hermite integration points

IOP Conf. Series: Materials Science and Engineering **836** (2020) 012002 doi:10.1088/1757-899X/836/1/012002

In order to reduce even more the computation efforts, an efficient truncation strategy is used, which aims to only retain the PCE coefficients associated to the low-order interactions [9]. Indeed, for engineering problems experiences have shown that high-order interactions have weak effect on the variability of the mechanical model. Let i be the maximum order of the retained interactions, according to An and Owen [10], the number of unknown coefficients of the PCE is $card(\mathcal{T}_n^{p,i})$, with $\mathcal{T}_n^{p,i} = \{\boldsymbol{\alpha} \in \mathbb{N}^n \text{ such that } |\boldsymbol{\alpha}| \leq p \text{ and } \sum_{k=1}^{n} \mathbb{1}_{\{\alpha_k > 0\}} < i\}$. It is clear that, for low values of the truncation maximum interaction order i, the *economy* defined as the ratio of $card(\mathcal{T}_n^{p,i})$ to $card(\mathcal{T}_n^{p})$ can be very significant. For instance, for $p = 8$, $n = 7$ and $i = 1$, the economy is around 10^{-2}. Once the unknown coefficients of the PCE are obtained, the statistical moments and the probability density function of the response of the mechanical model can be obtained by performing Monte-Carlo simulations on the polynomial approximation given by equation (1).

3. Application to fatigue crack growth

The uncertainty propagation presented in the previous section is applied now to conduct probabilistic analysis on Centre Cracked Plate (CCP) subjected to cyclic loads with a mean stress $\sigma_m = 15\ ksi$ and alternating stress $\sigma_a = 10\ ksi$. The CCP specimen, is made of 7075-T6 aluminium alloy, measures 12 inch wide (thus, w = 6 in.) by 35 inch tall and 0.09 inch thick and contains an initial crack 0.1 inch (thus, a = 0.05 in.) long located at its centre as depicted in figure 2.

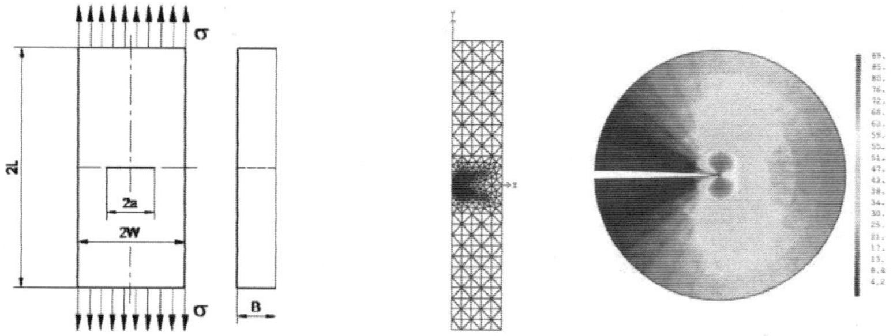

Figure 2. CCP specimen geometry (left), finite elements mesh (center), Von-Mises equivalent stress around the crack tip (right)

Although an analytical solution [11] is available for the Stress Intensity Factor (SIF) ΔK_I, as given by equation (3), a Finite Elements Model (FEM) of the CCP specimen is also developed using the software Cast3m [12]. This is in order to use an implicit mechanical model within the uncertainty propagation analysis which will be conducted later. Due to the geometry symmetry, only one half of the CCP specimen is modeled as shown in the finite elements mesh in figure 2. In addition, plane stress hypothesis is assumed since the specimen thickness is small compared to the two other dimensions of the specimen.

$$\Delta K_I = 2\sigma_a \frac{1 + 0.5\left(\frac{a}{w}\right) + 0.326\left(\frac{a}{w}\right)^2}{\left(1 - \left(\frac{a}{w}\right)\right)^{1/2}} (\pi a)^{1/2} \tag{3}$$

The obtained results, in terms of ΔK_I, for different loading conditions are given in table 1. As can be seen, numerical results are in good agreement with the reference solution since the relative error does not exceed 0.3 %. Furthermore, the accuracy of the finite elements model is not affected by the loading conditions. As shown in figure 2, based on the deformed mesh of the CCP specimen, the crack will propagate in opening fracture mode. In addition, linear elastic fracture behavior hypothesis is verified since the Von-Mises equivalent stress exceeds the yielding strength of the material $\sigma_{yld} = 75.9\ ksi$, only in the neighborhood of the crack tip. Since the accuracy of the FEM is now approved, it will be used in the sequel, firstly to fit the fatigue crack growth law and then to perform uncertainty propagation analysis based on PCE.

Table 1. Comparison of numerical and analytical results of the SIF range ΔK_I

σ_m (ksi)	$\sigma_a(ksi)$	Analytical ΔK_I	Numerical ΔK_I	|Error| (%)
5	5	5.6057	5.5934	0.2192
15	10	11.211	11.187	0.2192
15	3	3.3634	3.3560	0.2192

The fatigue crack growth is considered to be governed by the Walker law [13] which allows to take into account the effect of mean stress. Based on Walker law the relationship between the Fatigue Crack Growth Rate (FCGR) da/dN and the SIF range ΔK_I reads:

$$\frac{da}{dN} = C \frac{(\Delta K_I)^m}{(1-R)^{m(1-\gamma)}} \tag{4}$$

where a is the crack length, N is the number of loading cycles, $R = (\sigma_m - \sigma_a)/(\sigma_m + \sigma_a)$ is the stress ratio, C, m and γ are three real constants.

Table 2. Walker law parameters

m	γ	C	R^2_{LRGF}
3.82	0.56	8.19 10^{-10}	0.975

The parameters C, m and γ of Walker law are obtained through multiple linear regression analysis applied to da/dN and ΔK_I after transforming them to log-log space, and based on experimental tests performed by Hudson [14]. The estimates of Walker law parameters are given in table 2 and the fitted model is shown in figure 3(a). As can be seen, Walker law fits very well the experimental data since the goodness of fit parameter R^2_{LRGF} is close to 1. In addition, the fatigue crack growth life can be easily derived by integration of the Walker law as depicted in figure 3(b).

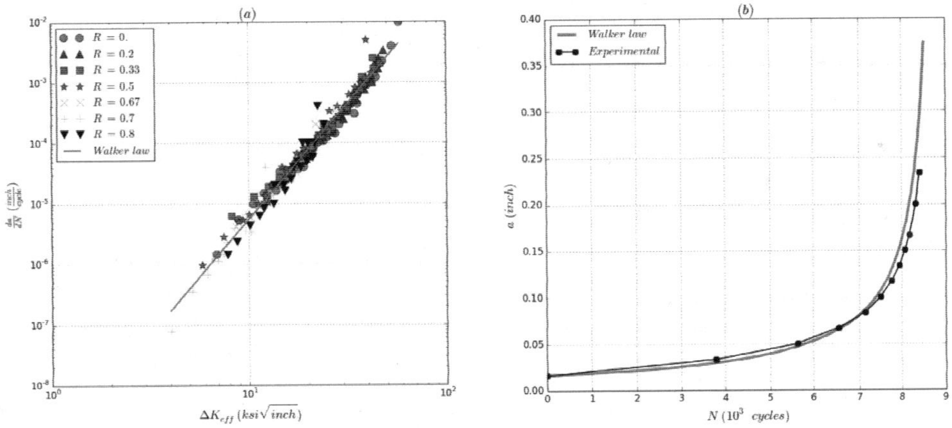

Figure 3. (a) Walker law fit to CCP specimen experimental data, (b) fatigue crack growth life based on the integration of Walker law

Statistical analysis of Hudson experimental data has shown that the parameters of Walker law are not deterministic. Determination of the statistical characteristics of these parameters is done using multiple linear regression on the experimental data and based of the Walker law rewritten as:

$$y = B + A_1 x_1 + A_2 x_2 \tag{5}$$

where $y = \log(da/dN)$, $B = \log(C)$, $A_1 = m$, $A_2 = -m(1-\gamma)$, $x_1 = \log(\Delta K_I)$, and $x_2 = \log(1-R)$. The statistical moments (i.e. the mean value μ and the variance σ^2), and the distribution type of the parameters A_1, A_2 and B of the modified Walker law are given on table 3.
The statistical characteristics of the parameters C, m and γ of the original Walker law are easily derived from the results given in table 3. Knowing this, uncertainty propagation analysis is performed

using PCE method on the previously developed finite elements model of the CCP specimen where the mechanical response is defined as the fatigue crack growth life N. First of all, the accuracy of the polynomial approximation to reproduce the real mechanical response is assessed. As can be seen from figures 4(a), 4(b) and 4(c), the polynomial approximation given by the PCE method fits very well the real mechanical behavior of the CCP specimen. Note that, 3rd order PCE is used to construct the polynomial approximation. The results show also that 3rd order interactions have no significant effect on the variability of the mechanical response since a good accuracy is achieved when only 2nd order interaction coefficients are considered in the PCE.

Table 3. Statistical characteristics of the parameters of the transformed Walker law

	Distribution	μ	σ^2
A_1	Normal	3.8681	2.2911 10^{-3}
A_2	Normal	-1.7052	3.5716 10^{-3}
B	Normal	-21.0737	26.4442 10^{-3}

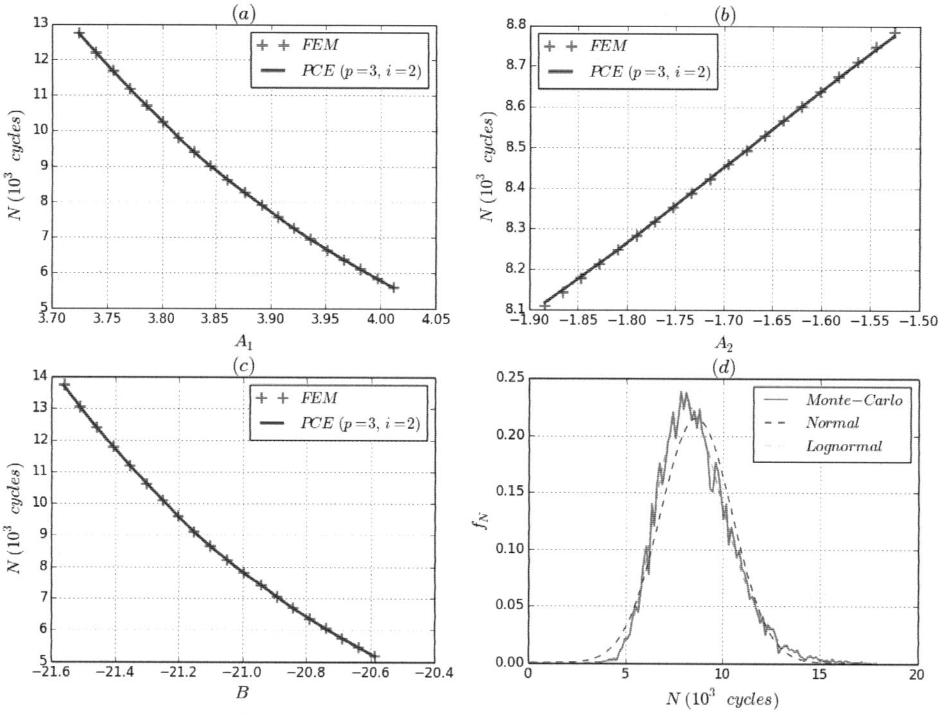

Figure 4. PCE approximation with respect to the uncertain parameters (a) $\boldsymbol{A_1}$, (b) $\boldsymbol{A_2}$, (c) $\boldsymbol{B}$, (d) PDF of the fatigue crack growth life

The two first statistical moments of the mechanical response (fatigue crack growth life N) are derived from the coefficients of the PCE as given in equation (1) and compared to the estimates given by *crude* Monte-Carlo Simulations (MCS) directly applied to the finite elements model and taken here as the reference solutions. As can be seen from table 4, the results obtained by the two methods are in good agreement since the relative error does not exceed 0.6%. The proposed method shows eventually its efficiency since only 40 FEM calls are required against 10000 FEM calls for *crude* MCS to achieve the same accuracy level on the estimates. In addition, the PDF of the fatigue crack growth life is constructed by performing MCS on the polynomial approximation given by PCE method. As can be seen from figure 4(d), the lognormal distribution fits it very well. Compared to the statistical moments, the PDF is quite informative since for each possible event we can attribute a probability of occurrence, which is very useful for engineers.

Table 4. Statistical moments of the mechanical response

	PCE	Crude MCS	|Error| (%)
μ	8633.73	8670.88	0.428
σ	1859.09	1869.26	0.544
FEM calls	40	10000	

4. Conclusion

In this paper an uncertainty propagation method based on PCE is presented to deal with fatigue crack growth problem. An efficient truncation scheme which allows to discard weak high order interaction effects is used to reduce the number of FEM calls to construct a polynomial approximation of the fatigue crack growth life. The remaining coefficients of the PCE are then computed using plan of experiments based on Gauss-Hermite integration points. The obtained results show that the variability of the fatigue crack growth life is very sensitive to the uncertainties associated to the material properties represented by the parameters of the Walker law since the coefficient of variation is around 21%. This information should be taken into account in the design of structures subjected to fatigue crack growth. Future works should focus on enhancing the proposed uncertainty propagation method and modelling uncertainties associated to geometrical parameters and loading conditions.

5. References

[1] Virkler DA, Hillberry BM and Goel PK 1979 *J. Eng. Mat. Tech* **101** 148.
[2] Bogdanoff JL and Kozin F 1985 *Probabilistic models of cumulative damage* (New York: Wiley– Interscience).
[3] Wu WF and Ni CC 2004 *Prob. Eng. Mech* **19** 247.
[4] Zhiping Qiu and Yuning Zheng 2017 *Int. J. Mech. Sciences* **134** 284.
[5] Wang L, Liang Y, Yang Y, Zheng Y 2018 *Theor. Appl. Fract. Mech.* **95** 104.
[6] Long XY, Liu K, Jiang C, Xia Y, Wu SC 2019 *Theor. Appl. Fract. Mech.* **103** 102268.
[7] Jakeman JD, Franzelin F, Narayan A, Eldred M, Plfüger D 2019 *Comput. Methods. Appl. Mech.Engrg* **351** 643.
[8] Ghanem RG and Spanos PD 1991 *Stochastic finite elements - A spectral approach* (Springer Verlag).
[9] Blatman G 2009 Adaptive sparse polynomial chaos expansions for uncertainty propagation and sensitivity analysis (PhD thesis Université Blaise Pascal, Clermont-Ferrand, France).
[10] An J and Owen A 2001 *J. Complexity* **17** 588.
[11] Tada H, Paris P and Irwin GR 2000 *The Stress Analysis of Cracks Handbook* (ASME press).
[12] CEA/LAMS – Saclay 1997 CASTEM 2000.
[13] Walker EK 1970 *The Effect of Stress Ratio During Crack Propagation and Fatigue for 2024-T3 and 7075-T6 Aluminum* In *STP462-EB Effects of Environment and Complex Load History on Fatigue Life*, ed. M. Rosenfeld, (West Conshohocken, PA: ASTM International) p 1.
[14] Hudson CM 1969 *Effect of Stress Ratio on Fatigue-Crack Growth in 7075-T6 and 2024-T3* Aluminum-*Alloy Specimens* NASA (Technical Note NASA TN D-5390, Langley Research Center, Langley Station, Hampton, VA).

IOP Conf. Series: Materials Science and Engineering **836** (2020) 012003 doi:10.1088/1757-899X/836/1/012003

Implementation of a quality framework on the launch phase of an automated assembly line for top class automotive chassis

Alessio Bertoldi, Francesco Gherardini*, Francesco Leali

Department of Engineering "Enzo Ferrari", University of Modena and Reggio Emilia, Via Pietro Vivarelli 10, 41125, Modena, Italy

*corresponding author: francesco.gherardini@unimore.it

Abstract. Today the assembly lines of cars chassis are highly automated by robotic operations. Even in the top class automotive sector, the production of aluminium chassis involves numerous automated operations, such as TIG and MIG welding, riveting and gluing. This practice allows, on the one hand, to reduce time and costs, improve process repeatability and quality standards. On the other hand, it requires the quality improvement of the whole process (from supplied parts approval to welding reworks minimization). The industrialization phase of a new car chassis and the launch of its automated assembly line are particularly critical, even more if the line has already been designed and only minimally modifiable. Therefore, this paper proposes the implementation of a quality framework to manage the launch of an automated assembly line of a new aluminium chassis of top class cars, selected as a case study. The framework was implemented, aiming at improving the entire process quality, and finally validated by critically comparing the results obtained with those relating to models currently in production. Due to their importance to the final quality, we focused on the welding operations, which require actions both on process parameters and supplied parts approval (e.g. tolerances on parts end cuts). The new line shows a clear improvement compared to the past, with highly significant reduction of welding non-conformances, high quality level and lack of many critical issues of the previous lines thanks to corrective actions taken in the early process stages, during the pilot phase.

Keywords. Quality management, Assembly process, Process FMEA, Car chassis, Geometric and tolerance specification, Robotic welding.

1. Introduction

A car chassis is commonly produced by assembling extrusions, castings, sheet metal parts through manual and robotic operations, such as TIG and MIG welding, riveting and gluing. Over the past few decades, the use of robotic operations has increased and, nowadays, even top class automotive chassis are manufactured in automated assembly lines, using lightweight alloy materials [1],[2],[3], such as aluminium. The required high quality standards are achieved thanks to the use of robots, able to guarantee an excellent repeatability of the process [4], with a consequent reduction of time and costs. On the contrary, automated operations are less flexible than manual ones, so the quality of the entire process must be evaluated and systematically managed, in order to limit errors, welding reworks, manual adjustments of parts. In particular, the final quality of the chassis is influenced by two main critical issues: firstly, the geometry and tolerances of the parts to be joined [5],[6], which cause gap variations between the parts, and secondly, the distortions due to the welding process [7]. The automotive fields, as well the other industrial ones, may benefit from the use of Computer-Aided technologies [8],[9],[10] that may improve the product/process quality and, apart from the design phase, can also be employed for dedicated prediction and analysis purposes [11],[12],[13]. However, the real process may differ from simulation due to simplification and uncontrollable variables. Therefore, the industrialization phase of a new car chassis and the launch of its automated assembly line are still particularly critical, so a quality management process is required and implemented through specific tools, such as the Advance Process Quality Planning (APQP) [14], endorsed by the Automotive Industry Action Group (AIAG). This is even more important if the line has already been designed and only minimally modifiable: this is the case of small batches assembled in previously existing lines (or partially redesigned) or, again, of outsourced lines, i.e. lines designed by the automotive company but commissioned to third-party suppliers, who are responsible for achieving the final quality of the chassis. Therefore, even in those cases, is it possible to apply specific tools and recover in quality?

This paper aims to address this question, by implementing a quality framework on the launch of an automated assembly line of aluminium chassis for top class cars. Using this approach, we can identified and promoted corrective actions in advance, in order to reduce time and variable costs related to the chassis production, to limit reworks, to minimize waste. Additionally, we reduced the plant surface necessary for the production line (i.e. the on-line safety buffer), which is a very critical variable in the analysed context.

The paper presents the quality framework based on APQP and systematic tools in Section 2. Then, Section 3 describes its application on the case study in and some final comments close the paper in Section 4.

2. Method

Focusing on the production launch phase of an already designed process in the automotive field, we propose a quality framework based on a "chunk" of the APQP, which is extracted and applied to the launch phase, highlighting the contributions of detection and correction tools in accordance with Fig. 1. The APQP is a framework for producing a product quality plan that will support the development of a product [15]. It formalizes a correct and efficient planning of the production (time plan), based on the results of product and process verifications, in order to guarantee a serial product that meets predetermined quality goals, by reducing times and costs.

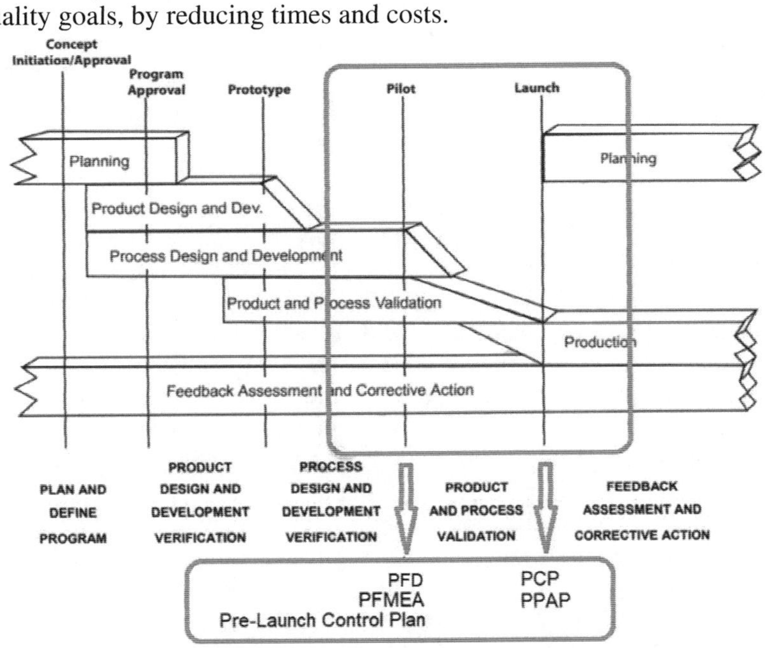

Figure 1. The implemented quality framework is highlighted in green, based on the "Process Design and Development" and the "Process Validation" phases of the APQP.

The Process Failure Mode and Effect Analysis (P-FMEA) is a technique for analysing the potential failure modes generated in the serial production process, by identifying the potential causes and effects

in order to establish corrective actions. The P-FMEA is implemented by a multidisciplinary product development team, capable of identifying potential process problems in advance, thanks to experience and know-how. The causes of failure modes are mainly due to the product/process development phase, thus much earlier than the Start Of Production (SOP). Therefore, their early identification leads to simpler and less expensive corrective actions. More generally, the FMEA is a structured method to analyse the entire internal production process. However the identification of failure modes has to be performed on each operation that constitutes the entire production process. This purpose is fulfilled by the Process Flow Diagram (PFD), a document that explains in detail the various phases defined by the line/process designer. In this way, each failure mode is analysed, defining the related causes, effects and risks, and associating an occurrence ranking, a severity ranking and a detection ranking, which allow to calculate a Risk Priority Number. This priority rule defines where to spent efforts (and investments) for facing failure modes with greater risk. In this way, detection and corrective actions can be planned, which may consist of a process change if the PFMEA is made in the initial phase of industrialization (prototype), or in an additional controls or scheduled maintenance activities. The FMEA is therefore the document that provides inputs for the activity of continuous improvement, involving different company departments, such as quality, maintenance, process technologies, production, technical department, or human resources, depending on the activity to be implemented. Moreover, the P-FMEA highlights the preliminary critical elements emerged before serial phases (the so-called *lessons learned*), which allow to define an effective control plan for serial production, with fine-tuning actions or additional controls.
The Production Part Approval Process (PPAP) is a formal and standardized method [16], which serves to test and approve a priori the production process of the supplier and of the related parts. The practice requires the supplier to submit the Part Submission Warrant (PSW) form during the pilot phase in order to obtain the approval, as well as a very specific report to document the parts conformity.
The Process Control Plan (PCP) is a serial production control tool capable to concentrate resources on the controls of the most critical features. The assessment of the characteristics to be checked and the relative control frequency are defined on the basis of: experience; results coming from the pilot phase; statistical trend of production indicators as Cp and Cpk indicators; severity of possible damage to the customer; key features defined by the customer.

3. Application to the case study

An automated assembly line for automotive chassis had already been designed, but not yet tested, as this activity had to be carried out throughout the pilot phase.
Scenario: The layout of the assembly line presents a main flow with different stations that generate many sub-assemblies of the chassis, then joined together into two main modules, called front extended frame and rear extended frame. These modules are welded together with other parts, such as the sills and sheet metal parts. The upper frame is generated and then completed by assembling the engine sub-frame and the rear sub-frame to the main body. Conversely, the previous assembly lines consisted of a much higher number of single parts directly welded together in two macro sub-assemblies, with consequent thermal distortions that must be compensated by appropriately locating the single parts in the welding jigs. Again, the new assembly line has a high percentage of automated operations (67% automated, 33% manual), necessary to cope with the high expected production volumes of the cars here produced. The previous lines are dependent on a marked manual contribution (e.g. in line A: 46% automated, 54% manual).
Critical elements: Analysing the statistics of the previous A, B, C lines, the Welding Non-Conformances (WNC) represent 23% of the total defects found at End of Line. WNC are not the most frequent defects, but the importance of the welding quality inside a chassis is high, as well as the cost necessary to welding reworks. In particular, two WNC have primary importance:
1) WNC due to end cuts of the parts to be welded, classified as weld burn-through (excessive penetration) and uncompleted welding, in the extrusion-casting joints: the main cause is the use of raw profiles for castings. The geometric tolerance of profile of a surface on the end cuts of castings is very large (around 2.5 mm), while on the end cuts of extrusions they are around 0.5-1 mm. Considering the precision of the torch of the welding robot around 0.01 mm, and the wire diameter around 1.2 mm, the geometric variability related to the prescription of raw profiles for castings is not compatible with the requirements

of precision of the welding robot. Conversely, in the extrusion-extrusion joints, the precision is higher since the tolerances are narrower, therefore the frequency of WNC caused by the variability of the end cuts is smaller.
2) WNC due to post-cooling deformations, caused by a high number of joints welded simultaneously in the same station. Therefore, the chassis distortion is not very predictable and, despite the adopted precautions, it affects the final tolerances on the chassis assembly. A greater modularity in the assembly process would make post-cooling deformations less incident. The analyses of the previous lines together with the company's experience have the task of highlight the critical elements before the launch phase.
Implementation: Various tools are concurrently implemented within the quality framework. In accordance with the "Process Design and Development" phase of the APQP, firstly the PFC and PFMEA tools are performed, whose results act as inputs for the subsequent tools. Then, in accordance with the APQP "Process Validation" phase, the PCP and PPAP tools are applied to monitor the production quality. The APQP provides for the preparation of a table showing the project time-plan shared with the customer during the planning phases. It allows to manage unexpected events, respecting the milestones.
P-FMEA: In accordance with ISO/TS 16949, the identification of fault modes must be performed on each operation of the entire production process, thanks to the Process Flow Diagram (PFD), a document that explains in detail the various phases defined by the line designer. It is necessary to draw up a detailed and accurate PFD (Fig. 2), with all the information needed to identify the failure modes (i.e. about the production process, the product, statistics related to similar models, etc.), their priorities and risks.

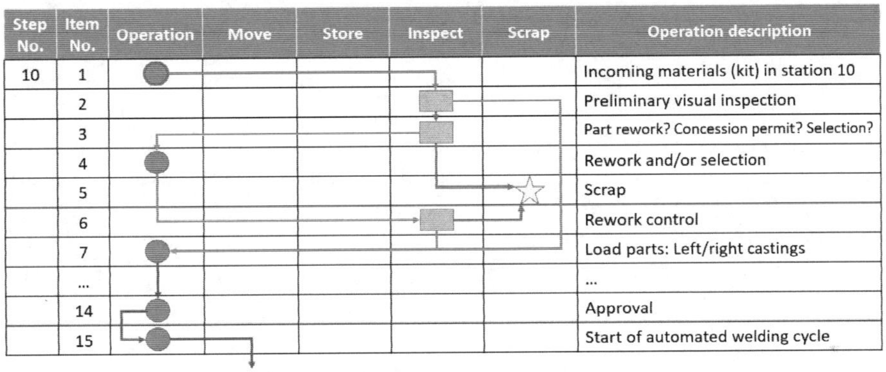

Figure 2. An extract from the PFD related to the automated welding station No.10.

In this paper, we focus on the effects of Non-Conformances (NC) of parts and process characteristics on welding operations:
a) Approval of parts from suppliers: In the previous lines, the parts were checked by the line operator, without a real inspection before the storing. This "free pass" approval has contributed significantly to produce chassis containing NC on the supplied parts, which often required subsequent reworking. Therefore, the P-FMEA is applied to identify recommended actions in parts approval (Fig. 3)
Based on the sequence formalized in the PFD, we analyse one operation after the other trying to understand for each identified failure mode, the cause, the generated effect and the current methods of prevention and detectability. A risk value is then assigned on the basis of the results of each analysis, which makes it possible to establish a priority in the possible assignment of a corrective measure.
Regarding casting parts, we redesign the joints with tighter tolerances, which have been rethought with single component processes in order to increase the level of precision of the parts to be welded, thus reducing the negative contribution that the geometric variability had on the quality of the weld. Therefore, we implement a control cycle in materials approval, thus guaranteeing the 100% of compliant parts in the line, at least in the pilot and serial launch production. Regarding extrusions, we have instead implemented a simple check of the packaging board, or rather of the quantity and arrangement of the parts as well as the conformity of the packaging, during the storage phase.

PROCESS PHASE	POTENTIAL FAILURE MODE	POTENTIAL EFFECT(S) OF FAILURE	POTENTIAL CAUSE(S) OF FAILURE	ACTUAL STATUS						RECOMMENDED ACTION(S)	RESPONSIBILITY & TARGET COMPLETION DATE	ACTION TAKEN	#OCC	#SEV	# DET	#RPN
				CURRENT PROCESS CONTROL PREVENTION	CURRENT PROCESS CONTROL DETECTION	#OCC	# SEV	# DET	#RPN							
INCOMING MATERIALS APPROVAL: EXTRUSIONS, CASTINGS, SHEET METALS	DIRTY PART	WELDING NON CONFORMITIES	LACK OF STORAGE SPACE	FOR CASTINGS: PACKAGING DATASHEET NOT PROVIDED	VISUAL CONTROL BY THE OPERATOR	10	5	7	350	PROVIDE CONTAINERS WITH CLOSED PACKAGING OR STORAGE IN A COVERED AREA	TECHNICAL OFFICE, QUALITY		1	5	7	35
	WRONG IDENTIFICATION	ASSEMBLY OF WRONG VERSION OF PARTS	WRONG IDENTIFICATION OF CONTAINER	NO PREVENTION	WAREHOUSE AUDIT	7	5	7	245	AUTOMATIC ERROR PROOFING ON EACH PART	TECHNICAL OFFICE, QUALITY	DATAMATRIX ON EACH PART	7	5	2	70
	WRONG QUANTITY	SUPPLY PROBLEMS	PACKAGING DATASHEET NOT COMPLYING BY SUPPLIER	NO PREVENTION	WAREHOUSE AUDIT AND STOCK CONTROL	10	2	9	180	CREATE A MATERIALS APPROVAL PHASE. DEFINE PACKAGING DATASHEET	LOGISTICS, QUALITY ASSURANCE	INCOMING MATERIALS CONTROL	4	2	7	56

Figure 3. Extract from the P-FMEA report of the new process, related to the materials approval phase: it shows only the failure modes with high Risk Priority Number.

b) Automated welding: The first step of the automated welding process is the manual or automated loading of parts and sub-assemblies to be joined. Based on statistics relating to the history of WNC, the assembly of an incorrect version of a part represents a rather risky mode of failure. Moreover, its possible rework entails a high cost also in terms of time. For this reason, we implement a datamatrix reading in the operator data sheet (Fig. 4). This reading is connected with the automated welding line software, so that it does not allow the operation if the operator has mistakenly loaded the wrong version of a part. The operator, in accordance with the instructions of each station represented in the operator data sheet, is then required to inspect the jigs locating the components and subgroup to look for any WNC.

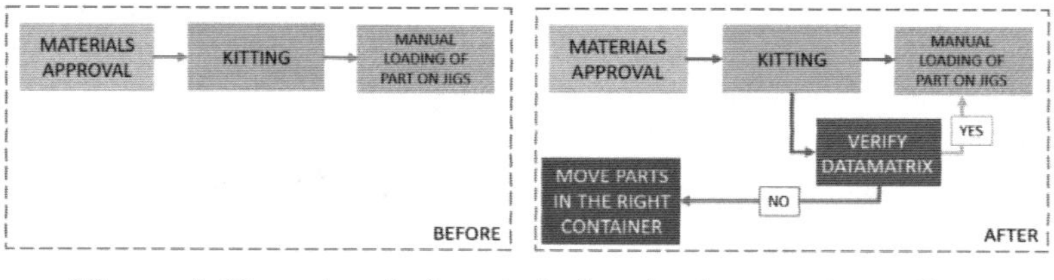

Figure 4. Flow chart before and after the datamatrix reading.

The next step was to consider the potential execution of an incomplete welding and of burn-through (excessively penetrated) welding. The case of burn-through welding is considered in Fig. 5.

PROCESS PHASE	POTENTIAL FAILURE MODE	POTENTIAL EFFECT(S) OF FAILURE	POTENTIAL CAUSE(S) OF FAILURE	ACTUAL STATUS						RECOMMENDED ACTION(S)	RESPONSIBILITY & TARGET COMPLETION DATE	ACTION TAKEN	#OCC	# SEV	# DET	#RPN
				CURRENT PROCESS CONTROL PREVENTION	CURRENT PROCESS CONTROL DETECTION	#OCC	# SEV	# DET	#RPN							
	BURN-THROUGH WELDING	WELDING NC: JOINT STRUCTURAL FAILURE	WELDING TORCH OUT OF GEOMETRY	NO PREVENTION	WDM AND VISUAL CONTROL 100%	4	7	7	196	TOOL CHECK PROCEDURE AND WDM MAPPING			2	7	3	42
	BURN-THROUGH WELDING	WELDING NC: JOINT STRUCTURAL FAILURE	PARTS OUT OF TOLERANCE	DIMENSIONAL CONTROL (INCOMING MATERIALS)	WDM AND VISUAL CONTROL 100%	4	7	7	196	COMPLETE WDM MAPPING			4	7	3	84

Figure 5. Extract from the P-FMEA report of the new process, related to operations with burn-through: it shows only the failure modes with high Risk Priority Number.

The characteristics influencing the temperature increase are: proximity of the torch, excessive gap of the joint to be welded that leads to a higher temperature peak in the extremal area of the two end-cuts, incorrect or unsuitable welding parameters. Through these actions, we reduce the on-line safety buffers, minimizing the plant surface necessary for the production line, which is considered a critical variable in the analysed context. The P-FMEA provides useful inputs to process the PCP.
PCP: Based on the collected data, improvement activities are implemented to ensure greater efficiency for future production. Other NC not yet found during the line fine tuning, emerge during the production process, from materials approval to the customer's approval. The elaboration of the PCP requires a multifunctional team that provides information from each company department to make effective the control action applied to the production process.
Fig. 6 shows part of the PCP for the new line. In particular, it lists the controls carried out during the preparation of the feeding parts for the assembly line (kitting) and for a robotic welding stations. The final document of the PCP does not specify in detail the characteristics to control, because the detailed control actions will be fully specified in other operational documents, e.g. check-lists.

Step No.	Process Name	Characteristics to control	Sample frequency	Control method	Reference doc.	References	Data registration	Reaction plan
05	Kitting	Identification	100%	Visual	BOM	Drawings		Stop/selection
90	Robotic welding	Datamatrix identification	100%	Datamatrix reader	Operation SOP	Operation datasheet	software	Isolate NC parts
		Preliminary inspection	100%	Visual				Out of line rework
		Parts loading	100%	Visual				
		SET-UP Approval		Visual		SET-UP procedure	SET-UP registration module	

Figure 6. PCP of the material preparation (kitting) and welding stations.

PPAP: The new line layout requires a supply approval process. The PPAP asks the supplier to report that all the specifications requested by the customer have been implemented into the final parts, e.g. the geometric conformity of parts, so as to minimize the problems related to parts supplied in the serial production. For this reason, the geometric report attached to the Part Submission Warrant (PSW) is examined, verifying the real accuracy of the parts in specific control points. The PSW examines the geometry of parts to evaluate the quality of the production process of this component. So, the Cp and Cpk indices are useful to express the productive capacity of a process in the form of a number (e.g. a $Cp > 1.33$ is required which indicates a deviation of about 96 ppm and a minimum value of $Cpk = 1.33$).

4. Discussion and conclusion

The new assembly line is mainly characterized by automated operations, so the reduced flexibility of the process cannot manage the same amount of WNC present in the previous lines. A quality framework based on the pilot and launch phases of the APQP was thus implemented, integrating quality tools and techniques (PPAP, PFMEA, PCP). Based on the experience gained during the pilot production, we acted from a technical point of view (chassis sub-modules, machined castings, etc.) and with appropriate methodologies applied in the launch phase. So, the following results are achieved:
- Management of the scheduled activities by means of the APQP, which allowed the Start of Production to obtain a production process cleaned of many critical issues.
- Preventive analysis of failure modes via P-FMEA, aimed at identifying as soon as possible the potential critical elements of the production process of a new chassis, before the launch phase. The P-FMEA allowed the company to analyse the production process step by step, evaluating and highlighting potential failure modes. These inputs led to important corrective actions: firstly, the creation of inspections in materials approval, which have contributed significantly to filtering NC in the supply, as well as new preventive maintenance procedures such as the periodic replacement of the wire guide. The impact of PFMEA is very important, especially with a view to continuous improvement, where it is also essential for cost containment to detect defects to be corrected in advance. From the *lessons learned*, the following corrective actions have been taken: a) increase the level of precision of the castings, which

have been redesigned and machined in the joint area to obtain joints with tighter tolerances, b) the adoption of a modular assembly sequence to reduce the thermal distortions caused by welding.
- The result in terms of quality of the final chassis is closely related to the quality of the supplied parts, whose compliance with the specifications of serial production was guaranteed by PPAP.
- Creation of a PCP plan before the start of serial production, aimed at maximizing the identification of WNC. We created 15 new control cycles related to the supplied parts.

Figure 7 shows the outcome of a quality assessment before and after the implementation of the quality framework during the pilot phase. Fig 7-a shows the quality performance of welded joints in a chassis assembled in one of the first runs of the pilot production. Fig. 7-b shows the welding quality performance obtained in a chassis at the end of the pilot production, thanks to the implemented quality framework. The comparison shows a reduction of WNC from 34% to a 1% relative to the last qualification performed, confirming the satisfactory result.

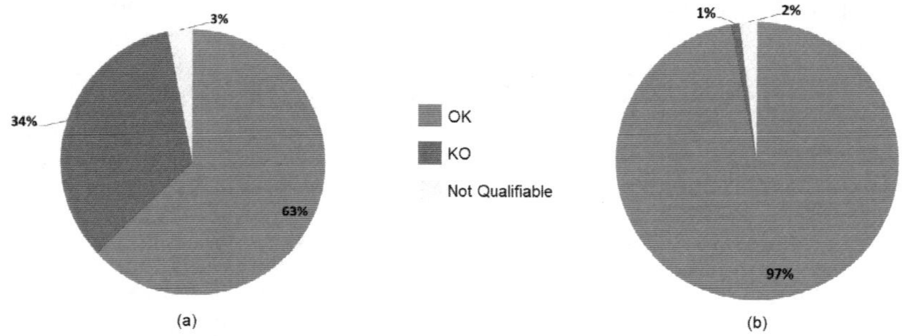

Figure 7. Quality of welded joints in (a) one of the first runs and (b) at the end of the pilot production.

Again, the quality of welded joints in the line A shows an average of 26% of WNC, calculated on the data throughout 2017. The new line shows an improvement even compared to the actual production.

In conclusion, the new line presents quality indicators that show a clear improvement compared to the past, thanks to a different global approach of quality management from pilot to launch phases. Future developments of this work will focus on the entire production, starting from the process line design.

References

[1] Cole G S, Sherman A M 1995 Light weight materials for automotive applications *Materials Characterization* 35(1) pp 3-9

[2] Rotondella V, Merulla A, Baldini A, and Mantovani S 2017 Dynamic Modal Correlation of an Automotive Rear Subframe, with Particular Reference to the Modelling of Welded Joints *Adv. in Acoustics and Vibration* Article ID 8572674.

[3] Mantovani S, Campo G A, Ferrari A, and Cavazzuti M 2018 Optimization Methodology for Automotive Chassis Design by Truss Frame: Preliminary Investigation sing the Lattice Approach *Proc. 25th ISPE Int. Conf. on Transdisciplinary Engineering*, IOS Press BV, Vol 7 pp 984-992

[4] Pini F, Panari D, Leali F 2019 Computer-aided integrated method for effective development of robotic welding processes of car chassis assembly *DYNA* 94(2) pp 150-154

[5] Barbero B R, Azcona J P and Pérez J G 2015 A tolerance analysis and optimization methodology. The combined use of 3D CAT, a dimensional hierarchization matrix and an optimization algorithm *Int. J. of Adv. Manuf. Technol.* 81(1-4) pp 371-385

[6] A, Gherardini F, Petruccioli A, Bonazzi E, Leali F 2020 Robust Parameter Analysis of Compliant Part Models for Computer Aided Tolerancing *Design Tools and Methods in Industrial Engineering. Proc. of the Int. Conf. on Design Tools and Methods in Industrial Engineering, ADM 2019*, September 9–10, 2019, Modena, Italy. Lecture Notes in Mechanical Engineering. Springer, Cham.

[7] Ruini F, Pini F, Leali F 2020 A CAE-based model of aluminium alloys welded T-Joints for TEP analysis *Design Tools and Methods in Industrial Engineering. Proc. of the Int. Conf. on Design Tools and Methods in Industrial Engineering, ADM 2019*, September 9–10, 2019, Modena, Italy. Lecture Notes in Mechanical Engineering. Springer, Cham.

[8] Cecil J, Kanchanapiboon A 2007 Virtual engineering approaches in product and process design *Int. J. Adv. Manuf. Technol.* 31(9–10) pp 846–856
[9] Mourtzis D, Doukas M, Bernidaki D 2014 Simulation in Manufacturing: Review and Challenges *Procedia CIRP* 25 pp 213-229
[10] Zardin B, Borghi M, Gherardini F, Zanasi N 2018 Modelling and simulation of a hydrostatic steering system for agricultural tractors *Energies* 11, 230
[11] King G S, Jones R P, Simner D 2003 A good practice model for implementation of computer-aided engineering analysis in product development J. *of Engineering Design* 14(3) pp 315-331
[12] Van Houten F J A M, Kimura F 2000 The Virtual Maintenance System: A Computer-Based Support Tool for Robust Design, Product Monitoring, Fault Diagnosis and Maintenance Planning *CIRP Annals* 49(1) pp 91-94
[13] Konstantinidis P A, Koltsakis G C, Stamatelos A M 1998. The Role of Computer Aided Engineering in the Design Optimization of Exhaust After-Treatment Systems *Proc. of the Institution of Mechanical Engineers, Part D: J. of Automobile Engineering* 212(3) pp 169-186
[14] Automotive Industry Action Group (AIAG) 2008 *Advanced Product Quality Planning (APQP) and Control Plan Reference Manual - Second Edition*
[15] Zhao Y 2011 *Information modeling for interoperable dimensional metrology* (Springer)
[16] Ford Motor Company 2008 *Global Phased PPAP Requirements Handbook Edition 2.0*

IOP Conf. Series: Materials Science and Engineering **836** (2020) 012004 doi:10.1088/1757-899X/836/1/012004

Decision support system for quality management of cosmetic products

O.V. Shabunina[1], A.I. Khalyasmaa[1], E.V. Korelina[1]

[1]Ural Federal University, 620000, Ekaterinburg, Russian Federation

shabunina-olga@yandex.ru

Abstract. The paper describes the structure and the mathematical core of the decision support system developed by the authors for assessing the quality of cosmetic products based on machine learning approaches. The main goal of the study was to improve the system of quality control of cosmetic products, in particular, the quality of perfumes and cosmetics using intelligent methods of data processing, formalization of knowledge and experience of the experts, thus providing automatization of the decision-making process. The practical-oriented goal of the study was to obtain knowledge-supported decisions regarding the quality of the cosmetic products using the developed system of quality assessment based on intelligent methods of data processing, providing the possibility of unsupervised learning and adjustment of the developed system when changing final product characteristics.

1. Introduction

The fashion industry, the enormous potential of the perfumery and cosmetics industry and the ideology of modern society, requiring a young and healthy appearance from a person, encourage people to use various cosmetic products. Therefore, today the list of cosmetic products, as well as their producers, is very extensive.

Quality control of the final products is an important function in quality management of any enterprise. It can be defined as a set of properties and characteristics that determine the feasibility of products to meet specific needs in accordance with the application sphere.

Each product has certain properties that characterize the quality level. General quality assessment criteria are established in regulatory documents: technical regulations, standards, technical conditions for specific types of products. It is generally known, that quality assessment of goods is associated with the optimal choice of the range of evaluation indicators, determination of their values by means of various methods and instruments and cross-checking with the basic values of the indicators. The general methodological approach to optimal criteria set determination and cross-checking of the indicator values is also well understood. However due to the specific features of cosmetic products associated with their functional and therapeutic properties, assessing the quality of such products is of great complexity.

2. Quality of cosmetic products and organization of its control

The quality of any product is to be assured at all manufacturing stages. The quality begins to play a crucial role at the formulation development stage in the course of scientific research activities. Then it is ensured during the production process and depends greatly on the quality of the ingoing materials, production process, methods and instruments of testing, storage and transportation. But before the start of mass production, the product is exposed to compliance assessment tests, which are carried out in

IOP Conf. Series: Materials Science and Engineering **836** (2020) 012004 doi:10.1088/1757-899X/836/1/012004

accordance with the established legislative requirements. Compliance is typically confirmed by the following certificates: Declaration of Conformity, Certificate of State Registration, Certificate of Conformity (depending on legal requirements for specific products). But it should be noted here that such a document is issued for a certain period of time and during the period of its validity it cannot guarantee the stability of the quality characteristics of all the manufactured products. Any production process is prone to technological failures. And the target consumer will easily stop using the product if any defect or quality violation is found. Consumer confidence is very easy to lose, and, subsequently, very difficult to regain. Therefore, one of the main goals of the manufacturing process is not to provide faulty goods.

Thus, the relevance of quality control is assured by up-to-date economy requirements and must be constantly monitored at all stages of the product life cycle.

Typically, to ensure the effective quality control process at the enterprises, quality control department and laboratory are established independently of other production units.

The quality control department, as a rule, includes a quality engineer and supervisors who carry out quality tests at the production sites. The laboratories include specialists who directly cross-check the samples using testing and measuring equipment.

Quality control department plays a crucial role in prevention of faulty goods production and is responsible for the monitoring results, preventing the delivery of faulty goods to the consumers. But, nevertheless, it should be noted that quality control department does not bear full responsibility for the quality of final products. This function is also allocated to the production workers. Therefore, all the personnel should be involved in quality monitoring. At workplaces the production personnel is obliged to monitor the proper operation of the equipment, the parameters of the technological processes, the conformity of products with reference samples at certain production stages and discard inappropriate products.

The main function of employees involved in the quality control process is to conduct tests and compare the results with specified (established) requirements, followed by determination of their compliance.

As a rule, the quality control process includes sampling at certain stages of the product life cycle, conducting established tests, recording test results. All recorded data is analyzed in order to obtain information about the possible malfunctions that could lead to quality degradation.

3. Decision support system for quality management of cosmetic products

To solve the above mentioned problems, the authors of the paper have developed a mathematical model of the decision-support system, providing quality assessment of the cosmetic products.

The main goal of this study was to improve the quality assessment system of the cosmetic products, in particular the quality of perfumes and cosmetics using intelligent methods of data processing, experts' knowledge and experience formalization approaches, providing automatization of the decision-making process. The practical-oriented goal of the study was to obtain knowledge-supported decisions regarding the quality of the cosmetic products using the developed system of quality assessment based on intelligent methods of data processing, providing the possibility of self-learning and self-adjustment of the developed system when changing final product characteristics.

The structure of the developed decision support quality control system of cosmetic products can be represented as the interaction of five separate sub-systems (Fig. 1):

- Data acquisition;
- Data generation;
- Knowledge discovery;
- Decision support;
- User interface.

IOP Conf. Series: Materials Science and Engineering **836** (2020) 012004 doi:10.1088/1757-899X/836/1/012004

At the data acquisition stage, all possible data sources are determined. This block is required to evaluate the amount of data, determine the order of its collection and the required storage volume in order to build the informative database at the next step. Data sources are given as follows:

- passport data for each ingredient from the particular supplier;
- testing protocols of quality control department;
- testing reports of microbiological laboratory;
- testing reports of the toxicology laboratory;
- quality control department final statement;
- consumer testing reports.

The data generation subsystem consists of two successive stages - data pre-processing and database development. The preliminary stage of data processing is of critical importance due to the need to scale and normalize data, as well as to exclude data outliers from the initial sample.
Based on the processed initial information about the object under consideration, a set of data stored in a certain ordered way (database) is generated. The following types of data can be stored in the database:

- physical and chemical indicators;
- microbiological indicators;
- toxicological safety indicators;
- organoleptic indicators;
- consumer testing results.

The sub-system of knowledge discovery is represented by two blocks - “intelligent data analysis” and “knowledge base”. Intelligent data analysis is understood as a sub-system of data processing based on machine learning, which gives the obtain to obtain aggregated differentiated estimation of cosmetic products\quality by using “knowledge base”. The “knowledge base” block is used not only for data storage, providing rules of interaction of various models (i.e. quality assessment of the final cosmetic product using the estimation results provided for ingredients), verification of existing knowledge, but also for generation the new ones.
In the course of data analysis, the quality assessment is carried out according to the principle “from particular to general”:

- state assessment of each particular ingredient;
- quality assessment of the cosmetic product in general.

Based on the obtained quality assessment results, a decision analysis is performed in the decision-support sub-system. The result in the form of ranked solution alternatives through the output sub-system goes to the user interface. The final choice is carried out by the decision maker.
As a result of quality control the following actions can be implemented:

- no actions required, the products comply to established requirements;
- quality assurance actions required, faulty goods are identified;
- product re-manufacturing with subsequent quality control;
- production cycle modification.

IOP Conf. Series: Materials Science and Engineering **836** (2020) 012004 doi:10.1088/1757-899X/836/1/012004

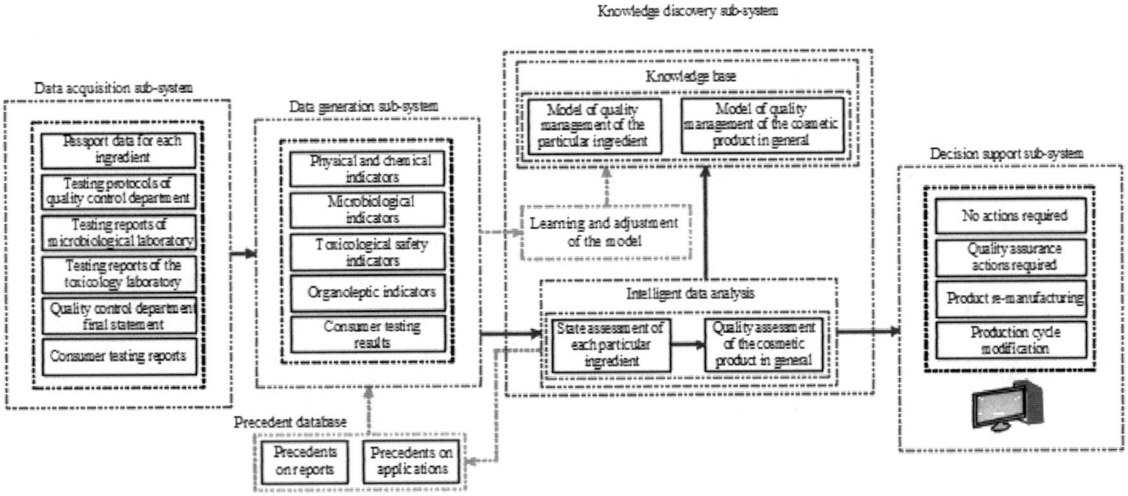

Figure 1. The structure of the developed decision support quality control system

4. Mathematical apparatus for cosmetic products quality assessment

In order to provide decision-support of quality control tasks, i.e. choosing the optimal solution based on the previous experience and rational analysis of all available data about the object under consideration, the instruments that implement human intelligence function are needed. Based on numerous author's studies, the Random Forest method was chosen as the optimal mathematical approach the introduced problem solution. In this case, Random Forest is used for the classification problem, arising from the multi-ingredient composition of the final cosmetic product, which is referred to as the supervised classification problem. The advantage of the proposed method is complete elimination of the overfitting problem even when the number of features significantly exceeds the number of observations. Moreover, the training sample required to compose a random forest may contain features measured on different scales: numerical, ordinal, and nominal, which is unacceptable for many other classifiers. A Random forest consists of a large number (ensemble) of decision trees.

The algorithm for constructing the Random Forest, consisting of N trees, if given as follows:

For each $n = 1, ..., N$: X_n sample is generated using the bootstrap process.

The decision tree b_n is built according to the X_n sample:

- according to the specified criterion, the best attribute is selected and the tree is partitioned until the sample is completely enumerated;
- the tree is built until each leaf node has no more than n_{min} objects or until we reach a certain tree height; - at each partition m random features are selected from n initial ones, and the optimal sample partitioning is searched only among them.

Then the final classifier given is as follows:

$$a(x) = \frac{1}{N}\sum\nolimits_{i=1}^{N} b_i(x). \qquad (1)$$

Thus, the decision is chosen by the majority vote. As a result, we get that Random Forest is bagging over decision trees, where the attributes are selected from some random subset of attributes during the training process.

In the Table below the example of cosmetic product quality control results is provided.

Table 1. Random Forest classification results.

Network data	Values
Number of features*	17
Number of training sample pairs	872
Number of test sample pairs	203
Average training error, % (Accuracy)	81,4
Average testing error, % (Accuracy)	**74,6**
Algorithm	Random-forest-classifier
Strategy for missing values	K-NEAREST-neighbours

In this table, the features are the ingredients such as plant extracts, essential oils, animal and / or vegetable fats, emulsifiers and co-emulsifiers, stabilizers, vitamins., The accuracy of identifying the quality of the developed system is 76% as shown in Table 2. This result confirms the effectiveness of the application of machine learning methods to solve these problems presented.

Conclusion

A brief overview of general consumer properties shows that, given the diverse purpose of cosmetic products, the range of properties varies significantly. Therefore, for a reliable quality assessment procedure of the cosmetic products it is necessary to conduct research on optimal range of quality indicators based on general methodological approaches. At present, the selection of consumer properties is often carried out without justification, choosing those indicators that are easy to evaluate. The main objective of the quality control system is to identify the stages at which problems may arise, and thus optimize the work of the personnel performing quality control: pay attention to where it is really needed and not to do unnecessary work where it is not required. The company considers the quality of its products as one of the most important indicators of its activities.

References

[1] Melnichenko T.A. Commodity research of perfumes and cosmetics. (for higher educational institutions). Rostov-on-Don: Phoenix, 2002. 288 p. (in Russian)

[2] Shepelev A.F. Commodity research and testing of perfumes and cosmetics. Rostov-on-Don: Publisher Center «March», 2001. 141 p. (in Russian)

[3] Shepelev A.F. Commodity research and testing of perfumes and cosmetics. Rostov-on-Don: Phoenix, 2002. 224 p. (in Russian)

[4] Yakovleva L.A., Kutakova G.S. Commodity research of perfumes and cosmetics. Saint Petersburg: Lan', 2001. 256 p. (in Russian)

[5] Cosmetic Chemistry by Florence Barrett-Hill, Ralph Hill. Paperback, 189 Pages, Published 2009 by Virtual Beauty Corporation ISBN-13: 978-0-473-12467-0, ISBN: 0-473-12467-X.

[6] Advanced Skin Analysis(2nd Edition) by Florence Barrett-Hill, Paul Dobbin, Jeff Burnett Paperback, 222 Pages, Published 2004 by Virtual Beauty Corporation ISBN-13: 978-0-476-00665-2, ISBN: 0-476-00665-1

[7] Cosmetic Dermatology, 2nd Edition edited by LESLIE BAUMANN Publisher: McGraw Hill, New York, NY; 2009

[8] Samuylova, L. I. Cosmetic chemistry in 2 parts, Part 1: Ingredients / L. I. Samuylova, T. A. Puchkova. Moscow: Cosmetic chemists' school, 2005. 386 p. (in Russian)

IOP Conf. Series: Materials Science and Engineering **836** (2020) 012005 doi:10.1088/1757-899X/836/1/012005

Digital twin technology as an instrument for increasing electrical equipment reliability

A.I. Khalyasmaa[1], S.A. Eroshenko[1], D.V. Shatunova[1], A. A. Larionova[1], A.O. Egorov[1]

[1]Ural Federal University, 620000, Ekaterinburg, Russian Federation

lkhalyasmaa@mail.ru

Abstract. The article describes the experience of practical realisation of digital twin technology for voltage and current measurement transformer 110 kV installed in real power station. In terms of the research electrical equipment state monitoring system has been developed as a set of sensors and diagnostic system, based on essentially different approaches of non-defective diagnostics. 3D model construction of the equipment under monitoring is also described as a part of digital twin concept. It is shown the valuable impact of digital twin technology application.

1. Introduction

Digital twin concept assumes to use a number of technical solutions, namely, set of sensors, monitoring, data acquisition and data storage systems, internet of things technologies for data transmitting and various data processing approaches and algorithms e.g. artificial intelligence.

Parametric virtual prototype of real object, system or process is resulted from digital twin technology application. Based on all historical data and current measurements it allows not only to simulate but to solve the problem of life cycle management.

Under the Russian Federation economics digitalisation overall and particularly digitalisation of the energy sector digital twin approach becomes new instrument for overcoming essential technological issues and achieving national goals of scientific and technological development of the industry in terms of integration of latest technologies in the field of smart managing systems, big data analysis, machine learning and artificial intelligence techniques and providing effective ways against anthropogenic challenges for society, economics and government.

2. Basic concepts and implementation principles

Initially digital twin concept was proposed by Michael W. Grieves in [4]. Nowadays, digital twin is determined as real time replica of all product life cycle components including interfaces, functionality and algorithms [5]. Digital twin combines data of real time operation, detailed mathematical model, which parameters are determined based on collected data. A long with digital twin digital shadow concept has been introduced.

Digital shadow can be determined as system of interconnections which describes real object behaviour usually in normal conditions. Such interconnections can be found in collected superfluous big data coming from real object via industrial internet technology

Digital shadow can forecast real object behaviour only for the states which data has been collected, but not for the new states [6].

In terms of [1] and [2] JSC Rosseti (major Russian electrical grid operator) developed the concept of digital transformation 2030 [3], which defined basic directions pf such activity. The concept determines destination model, Big Data management instruments, digital network framework, company digital management. Possible digital technologies integration effect has also been evaluated. New scientific and technological topics was formulated as follows [3]: digital substation, active-adaptive network with distributed smart system of automated control and management, complex business process effectiveness and management system automation, implementing of new technologies and materials in electrical industry, perspective technologies of digital transformation.
In terms of the concept digital twin technology applications have been determined as follows: utility connection, operation and contingency control, technical maintenance and repair, supervisory system of business asset and capital development. Business asset managing is closely connected with technical state estimation of the equipment and particularly with state analysis based on main functional blocks.
In this case digital twin technology which is considered as combination of physical and digital models allows not only to increase high voltage electrical equipment technical state identification accuracy but to establish hidden relations between various parameters and to collect new practical knowledge using Big Data analysis and machine learning algorithms. The general idea and further application of digital twin concept are depicted by Fig. 1.

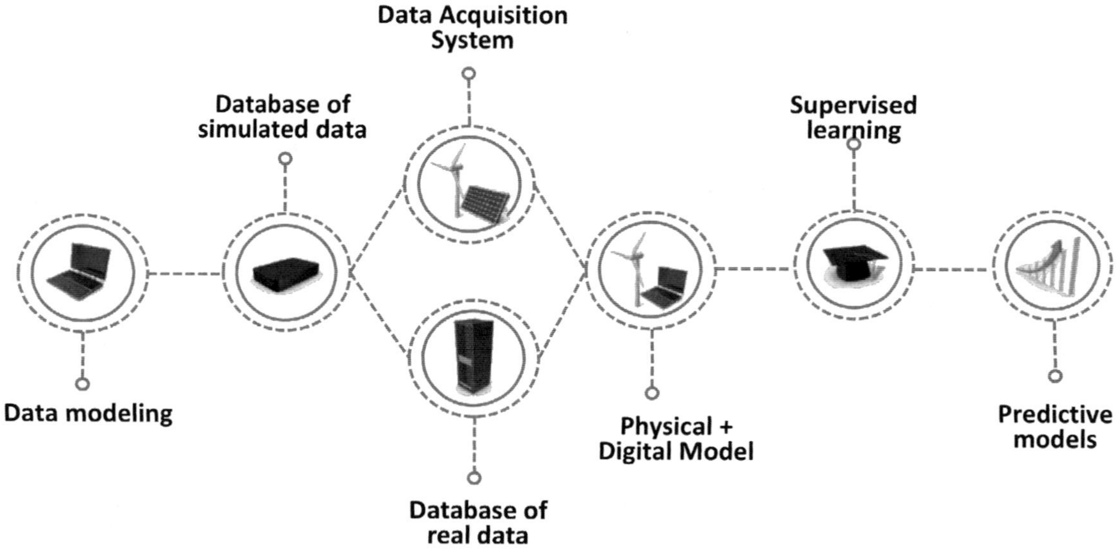

Figure 1. Digital twin constructing technology

3. Electrical equipment digital model realisation

Proposed technique was implemented for active oil-filled measurement current and voltage transformers installed in open electrical switchgear 110 kV of one of power stations in Russia.
Digital model was established by means of diagnostic system integration developed under the project and based on a number of non-destructive analysis approaches, namely ultraviolet light monitoring, heat monitoring and internal discharge measurements for oil-filled equipment isolation.
The complex system for technical diagnostic of measurement current and voltage transformers 110 kV has been established. The system consists of on-line monitoring system and portable units for technical state identification without switching equipment off, namely:

- Current and voltage transformer isolation internal discharge monitoring system;
- Current and voltage transformer monitoring and diagnostic system CDM delepoded by Dimirus: CDM unit, connection device DB-2/TT, measurement controlling device Pt-100A, moisture controlling device SHm-1;

- Stationary infrared light monitoring system, which includes thermal camera packed with objective and central server;
- Ultraviolet light monitoring system, consisting of Filin-6 defect detector.

When the diagnostic system was established equipment parameters measurements was performed, for each equipment was chosen its own set of parameters. Technical state analysis was performed based on collecting data. Groups of defects or limit states for various parts was defined using obtained data of different technical on-line diagnostic approaches. It should be mentioned that obtained limit states can not be identified by means of conventional diagnostic procedures at all or on so early defect development stage. Developed diagnostic system is depicted by Fig. 2.

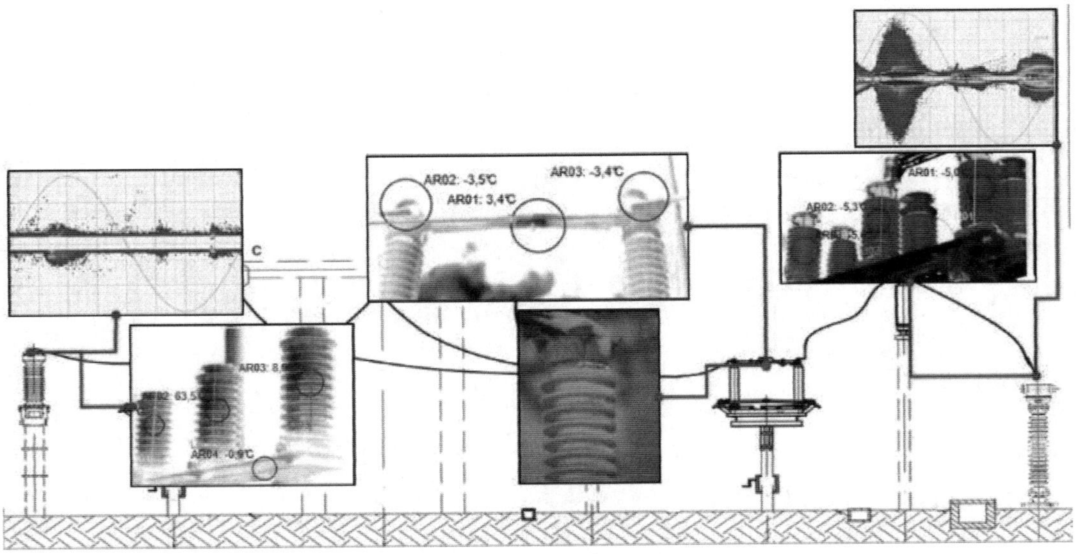

Figure 2. Electrical equipment digital twin diagnostic system realisation

4. Realisation of current and voltage transformer 3D model

One of the parts of digital twin concept is the development of electrical equipment 3D model. Such model can be used for calculation and analysis of operation. Under the research 3D models of current and voltage transformer 110 kV installed in open substation switchgear have also been developed. 3D model development was performed by means of inverse engineering problem which was used to restore unit size. Restoration process was supplied by electrical equipment scan-file because conductive part carried voltage and it was allowed to measure them directly. Proposed model was developed in AutoCAD with the scale 1:1. Obtained models are depicted by Fig. 3–4.

Figure 3. 3D model of current measurement transformer 110 kV (carcase)

Figure 4. 3D model of current measurement transformer 110 kV

5. Analysis of current and voltage transformers

Technical state estimation algorithm was performed using on of machine learning algorithm, namely decision tree gradient boosting. Indeed, various artificial intelligence approaches has been probed as a solution for the problem e.g. neural network [6], genetic algorithms [7], neural-fuzzy logic [8], machine learning [9]. Each method has its own advantages and disadvantages from result analysis efficiency of technical diagnostic and monitoring and in more detail, it is described in [10]. Gradient boosting like other boosting algorithm consequently build base model in such manner that each next model increases the quality of the group. Decision tree gradient boosting builds the model in form of tree sum [11]:

$$f(x) = h_0 + \upsilon \sum_{j=1}^{M} h_j(x) \tag{1}$$

where h_0 – first approximation, $\upsilon \in (0,1]$ – parameters, which regulates the learning speed and specific tree influence on the whole model, $h_j(x)$ – regression decision tress. New additive trees are added to the sum by mean of empiric risk greedy optimisation, described by some losses function $L(y, y') = L(y, f(x))$. Proposed method can be used for any differentiable losses function without valuable correction.

As a result, the algorithm is represented by a piecewise constant function describing the relation between the state Y (or defect, depending on the type analyzed data) and the set of operational parameters X_n. The average error Q in percentage [%], which was used to evaluate the testing error of the algorithm a on the sample X^l, is calculated as follows:

$$Q(a, X^l) = \frac{1}{l} \sum_{i=1}^{l} |a(x) \neq y^*(x)| \cdot 100\%, \tag{2}$$

where $|a(x) \neq y^*(x)|$ – error indicator; $y^*(x)$ – true value of the parameter; l – number of observations.

6. Conclusions

Digital twin technology allows not only to create completely new approach for electrical equipment state estimation in terms of new recommendations for every non-defective diagnostic technique implementation, but to from defect library for each technique to simplify and increase the accuracy of state identification.

Also, digital twin technology enables to stress equipment group with limit state of the equipment as a whole but also its specific parts. Such limit state cannot be determined by means of conventional diagnostics approaches established in technical standards at all or at least on so early stage of defect development.

References

[1] Michael W. Grieves Digital Twin: Manufacturing Excellence through Virtual Factory Replication LLC, 2014, 7 p.
[2] Tao F. et al. Digital twin-driven product design framework//International Journal of Production Research. 2018. C. 1-19.
[3] Kokorev D.S., Yurin A.A. Digital doubles: concept, types and benefits for business // Colloquium-journal No. 10 (34), 2019/Technical science. S. 31-35.
[4] Tao F. et al. Digital twin-driven product design framework //International Journal of Production Research. 2018. C. 1-19.
[5] Siemens, Siprotec-Digitaltwin [Электронный ресурс]: офиц. сайт. URL: http:// www.siemens.com/siprotec-digitaltwin Загл. с экрана (дата обращения 26.06.2019).
[6] Khalyasmaa, A.I., Dmitriev, S.A., Kokin, S.E. Monitoring and diagnostics systems application at 35-110 kV substations. Advanced Materials Research. Vol. 694 697, p. 1329-1333.

[7] Kokin, S., Dmitriev, S., Khalyasmaa, A. Evaluation model for urban power supply systems. Advanced Materials Research. Vol. 468-471, p. 1642-1648.
[8] Khalyasmaa, A.I., Dmitriev, S.A. Expert system for engineering assets' management of utility companies. Proceedings - SDEMPED 2015: IEEE 10th International Symposium on Diagnostics for Electrical Machines, Power Electronics and Drives. 7303724, p. 421-427.
[9] Khalyasmaa A.I., Eroshenko S.A. Data analytics platform for power equipment intelligent lifecycle management. D2-315. 47th CIGRE Session Proceedings, 2018.

Acknowledgments
The reported study was supported by Russian Science Foundation, research project No. 18-79-00201.

IOP Conf. Series: Materials Science and Engineering **836** (2020) 012006 doi:10.1088/1757-899X/836/1/012006

Silver Line Defect Reduction of Rear Lens Combination Lamp Process

Titiya Lapwattanamongkol[1] and Parames Chutima[1,2]

[1]Department of Industrial Engineering, Faculty of Engineering, Chulalongkorn University, Bangkok, Thailand 10330
[2]The Royal Society of Thailand

[1]pam.titiya@gmail.com, [2]parames.c@chula.ac.th

Abstract. The company in this research supplies the automotive lamps to many automotive manufacturing companies. The research focuses on the Rear Lens Combination Lamp Injection Process which causes a lot of silver line defectives and has the most impact on customer viewpoint. The DMAIC phases of the Six Sigma methodology are applied to solve the problem. New parameters are set at the optimal values as the result of the experimental design and analysis. After the improvement is implemented, the silver line defects are reduced from 10.21% to 2.41% (7.8% defect reduction)

1. Introduction

In order to improve customer satisfaction on the vehicle, the automotive manufacturing company has focused on part appearance quality improvement. The latest model is going to launch. After collecting of previous model defective ratio from suppliers [1,2], the rear combination lamp is the first rank defective. Rear combination lamp is consisted of 4 main components. There are 1. Lens, 2. Extension, 3. Reflector and 4. Housing. Almost of defectives are occurred on the lens component. Therefore, the rear lens combination lamp process is selected to study for quality improvement. There are 2 sub-processes which are lens injection process and surfacing process. Defects are mainly presented at lens injection process due to any defective is 100% to be scrapped and cannot be passed to surfacing process. From the historical data, the Silver line defective is the highest ratio from the rear lens combination lamp process.

The Six Sigma methodology is applied for this research to improve the process. DMAIC consists of 5 phases, i.e. Defining – identify the problem to be improved, Measurement – ensure the accuracy and precision of measurement system and identify causes of problem, Analysis – identify the significant factor that directly affect by causes of problem, Improvement – utilize design of experiment method to get the optimal value, Controlling – apply the optimal value from improvement phase and standardize to control the quality in the process [3].

2. Define phase

The process of the rear combination lamp is consisted of 2 sub-processes. The defect is mainly from Lens injection process [4]. The historical defect data from the previous model (recorded during April 2018 – March 2019) has shown in figure 1 that the highest defect came from Silver line.

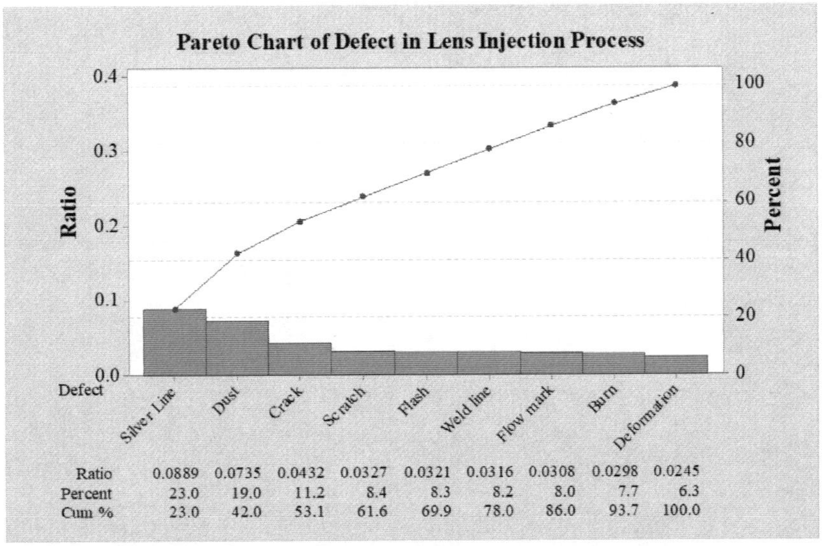

Figure 1. Pareto Chart of Defect in Lens Injection Process during April 2018 – March 2019.

3. Measurement phase

In this phase, 2 main methods are conducted those are *GR&R* (Gage Repeatability & Reproducibility) and *Problem Identification Analysis.*

GR&R is applied to be ensured the accuracy and precision of the measurement system. Measurement variation consists of 2 factors, repeatability –variation from equipment and reproducibility – variation from the inspector. Two inspectors have been provided with the training and assigned to be performed repeatedly 2 times with random inspection on 20 different critical of Silver line defect lens components. The result of this method, as shown in 'table 1', can be summarized that the accuracy and precision of the measurement system are acceptable.

Table 1. Result of Gage Repeatability & Reproducibility.

Index	Inspector 1	Inspector 2	Both Inspectors
%Repeatability	100%	100%	-
%Reproducibility	100%	100%	-
%Repeatability Effectiveness	-	-	100%
%Reproducibility Effectiveness	-	-	100%

Problem Identification Analysis consists of 2 steps to identify the factors affecting the silver line defect in the lens injection process. *Cause-and-Effect Matrix* is the first step to find out the root causes by working team. After listing up all possible factors, the working team has scored the relation of the cause-and-effect matrix. From descending sorted of relation scoring in figure 2, 10 from 17 factors are significantly highlighted by 80/20 rule from the Pareto chart.

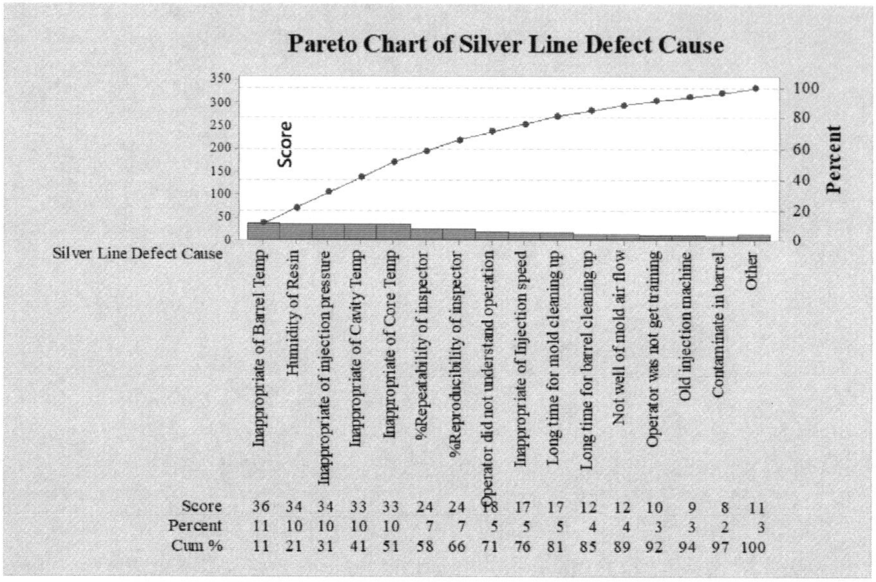

Figure 2. Pareto chart of Silver line defect causes.

Continue with the second step, these 10 factors are deeply analyzed by Failure Mode and Effect Analysis (FMEA) [5] and scoring based on RPN. Working team scores RPN by the level of Severity, Occurrence and Detection. The three highest factors from FMEA shown as figure 3, there are Cavity Temperature, Core Temperature and Barrel Temperature, are selected to Analysis phase.

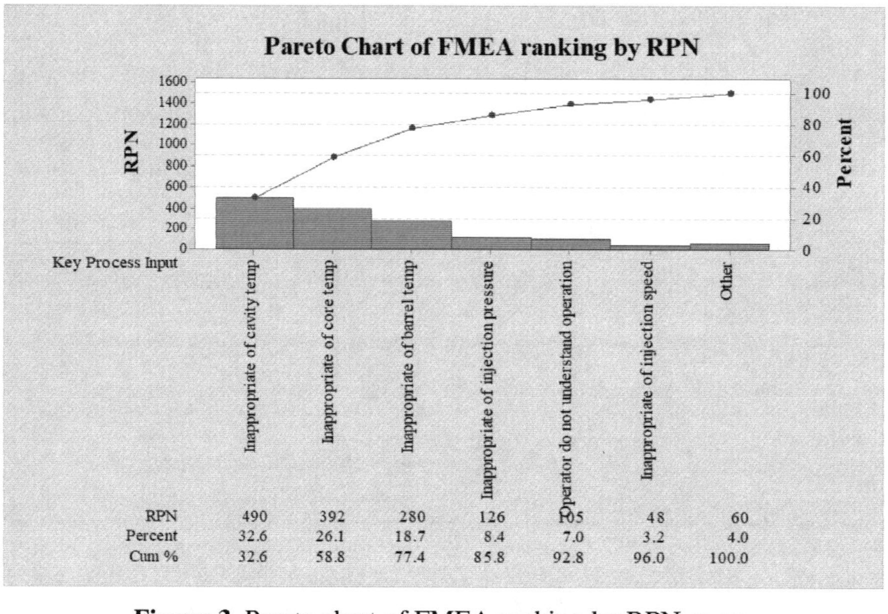

Figure 3. Pareto chart of FMEA ranking by RPN score.

4. Analysis phase

The three factors from Cause-and-Effect Matrix and FMEA, i.e. Cavity Temperature, Core Temperature and Barrel Temperature of the lens injection process, are analyzed to be ensured that they are the significant factors of Silver line defect by Hypothesis testing. Two-Proportion Testing is applied to this phase. Defect ratio of 2 Difference values parameters is collected to find the appropriate sample size from Minitab software. Then hypothesis testing at 95% confident interval is conducted as the result shown in 'Table 2'. All three factors' P-Value is less than 0.05. The null hypothesis is rejected. Mean that all three factors are affected by the Silver line defect in the lens injection process.

Table 2. Two-Proportion Testing of Three factors of the Silver line defect.

	Cavity Temp		**Core Temp**		**Barrel Temp**	
	88°C	**90°C**	**92°C**	**94°C**	**250°C**	**260°C**
Sample size	66		36		53	
No. of Defect	3	15	2	11	16	2
% of Defect	0.05	0.23	0.06	0.31	0.30	0.04
P-Value	0.002		0.004		0.000	

5. Improvement phase

From all three significant factors, the design of the experiment is conducted as full factorial design by 3 factors and 2 levels with 2 replicates as shown in 'Table 3'. The experiments resulted in Minitab software as shown in figure 4, the main effect Cavity, Core and Barrel temperature are significant at 95% confident interval.

Table 3. Lens injection process parameters and level.

No	**Input factor**	**Symbol**	**Level (°C)**	
			Low	High
1	Cavity Temperature	Cavity Temp	88	90
2	Core Temperature	Core Temp	92	94
3	Barrel Temperature	Barrel Temp	250	260

Analysis of Variance

Source	DF	Adj SS	Adj MS	F-Value	P-Value
Model	7	0.068282	0.009755	169.75	0.000
Linear	3	0.046113	0.015371	267.48	0.000
Cavity Temp	1	0.019924	0.019924	346.71	0.000
Core Temp	1	0.007114	0.007114	123.80	0.000
Barrel Temp	1	0.019074	0.019074	331.92	0.000
2-Way Interactions	3	0.020438	0.006813	118.55	0.000
Cavity Temp*Core Temp	1	0.002439	0.002439	42.45	0.000
Cavity Temp*Barrel Temp	1	0.015915	0.015915	276.94	0.000
Core Temp*Barrel Temp	1	0.002084	0.002084	36.27	0.000
3-Way Interactions	1	0.001731	0.001731	30.13	0.001
Cavity Temp*Core Temp*Barrel Temp	1	0.001731	0.001731	30.13	0.001
Error	8	0.000460	0.000057		
Total	15	0.068742			

Figure 4. ANOVA result of Lens injection process.

The normal probability plot of the residual for defects represented the errors are normal distribution as the straight line. The residual versus fit represented the variance stability. The residual versus order represented the independent. All are shown as figure 5, that means the experiment has followed the condition of the experiment design NID $(0, \sigma^2)$. Therefore, the optimum value of each factor should be 88°C of Cavity Temperature, 92°C of Core Temperature and 260°C of Barrel Temperature.

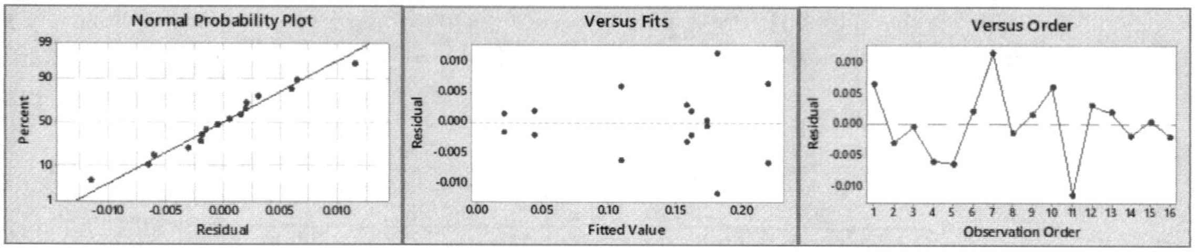

Figure 5. Residual Plot for Defects.

6. Control phase

The result of the Improvement phase is implemented in the lens injection process. The defective data comparing before and after improvement shown as figure 6 Running the operation for 8 days. Silver line defect ratio has been reduced for 7.8 %

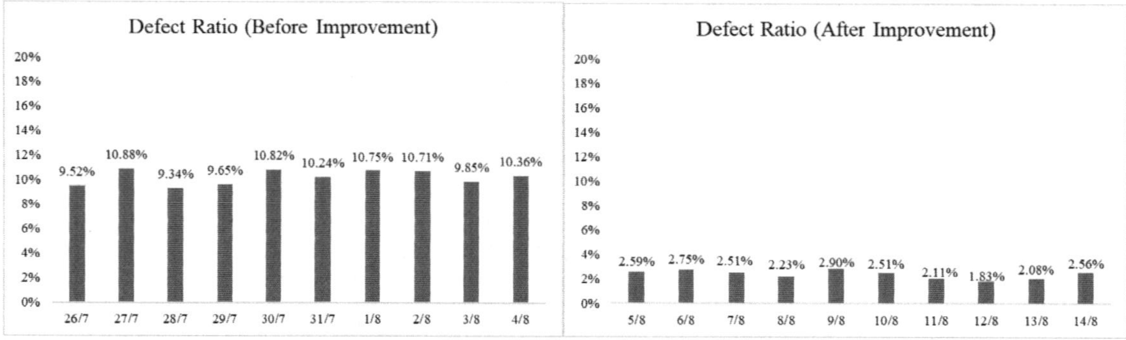

Figure 6. Defect Ratio Comparison between before and after improvement.

After the improvement, Standard Operation must be updated and revised to maintain the quality of lens injection process, i.e. 1. The appropriate value of cavity, core, and barrel temperature for this product (specific to the type of resin) 2. Add the working step of temperature control checking by the operator [6]. In addition, both operation and inspection training have to be given to operators and inspectors. Post-test is recommended to conduct after training and periodically re-training every 6 months to be maintained their skill.

7. Summary

The objective of the research is to reduce the silver line defects in the lens injection process by applying the Six Sigma methodology. The DMAIC is applied step-by-step to analyse the root causes and to indicate the new optimal parameters. The result reveals that after implementing the new parameter settings to the process, the silver line defects are reduced by 7.8%.

8. References

[1] Douglas C Montgomery, 2005. *Design and Analysis of Experiments*. USA: John Wiley & Sons, p. 235 - 277

[2] Coronado, R.B. and Antony, J., 2002. *Critical success factors for the successful implementation of six sigma projects in organizations*, p. 92 – 99

[3] Harry, M.J. and Schroedor, R. 2000. *Six-Sigma-The Breakthrough Management Strategy Revolutionizing the world's top corporations*. New York: Currency and Doubleday, p. 138 – 230

[4] Kalamdani, R. and Khalaf, F. 2006. *Application of Design for Six Sigma to manufacturing process design at Ford PTO. Int. J. of Product Development*. 3, ¾: p. 369 – 387

[5] Park, S.H. 2003. *Six Sigma for Quality and Productivity Promotion*. Tokyo: Asian Productivity Organization, p. 113 – 121

[6] Schroeder, R.G., Linderman, K. Liedtke, C. & Choo, A.S., 2008. *Six Sigma: Definition and underlying theory*. Journal of Operations Management, Volume 26, p. 536-554

IOP Conf. Series: Materials Science and Engineering **836** (2020) 012007 doi:10.1088/1757-899X/836/1/012007

Electrical energy storage systems for increasing technical and economical characteristics of gas engine power plants

S.A. Eroshenko[1], V. D. Melnikov[2], G. B. Nesterenko[2,3], V. M. Zyryanov[2],

[1] Ural Federal University, 620002, Ekaterinburg, Russian Federation
[2] Novosibirsk State Technical University, Novosibirsk, 630073, Russian Federation
[3] Energy Storage Systems LLC, Novosibirsk, 630007, Russian Federation

nesterenkogb@yandex.ru

Abstract. Developing of new technologies such as electrical energy storage systems (EESS) opens new ways to increasing efficiency, decreasing exhaust waste and fuel consumption in traditional fossil fuels power station. This paper contains actual testing results of electrical energy storage system based on lithium-ion technology working with gas engine generator and altering load power. EESS parameters: rated power 100 kVA, rated capacity 153 kWh. As is known, gas generators are sensitive to changes in load power, obtained results show the effectiveness of the usage EESS to supply sustainability of generators in high amplitude and frequency altering load power environment. These results achieved through the ability of the developed converter and control strategy to compensate rapid deviations of the load power. In additional, in control system realized others common functionality: active power limitation, voltage and frequency control, reactive power control, power factor control and others. With the usage EESS, the nominal power of plane station should be lower than without EESS.

1. Introduction

The global interest in reducing the volume of hydrocarbons burned caused a spread interest in the using of renewable generation (mainly solar power and wind power) at the electric power industries of most developed countries. However, this type of generation is not able to completely displace the traditional power stations because of its operation peculiarities, namely of the stochasticity of the generated power. This peculiarity allowed the development of a whole class of devices, which are now commonly referred to as electrical energy storage systems. Also, the development of electrical energy storage systems is due to the spread of electronic portable devices and electric transport, these industries have increased the specific capacity of batteries and reduce the cost of their production (the cost of batteries can be more than 50% in the cost of the electrical energy storage system).

In the world practice, electrical energy storage systems are used not only for supporting RES, but also, for frequency and voltage regulation, electric supply reserve etc. However, statistics show that the global production of electricity is provided mainly by burning hydrocarbons [1], and given their low cost, established system of extraction, preparation, transportation, as well as a high-level development of technologies for obtaining electricity from them, these energy sources, remain the main source of electricity. At the moment it is too early to discuss about large-scale application of electrical energy storage systems to increase the efficiency of electricity generation, which is associated with their high cost, however the application in local isolated power systems is a promising direction and it possible to improve the efficiency of traditional power plants.

2. Features of Isolated Power

About 2/3 of the territory of Russia does not have a centralized power supply, (Figure 1). There are thousands of settlements, industrial enterprises, as well as oil and gas production facilities at this territory. To provide power supply for these facilities mainly diesel or gas local power stations are organized. Often, these areas are characterized by poor transport access. Fuel delivery to facilities, especially located in the Far North and the Far East, involve high costs. Moreover, it is necessary to store a large amount of fuel in the winter when there is a transport link with large settlements.

Figure 1. Electrified of the Russian Federation

When designing diesel power plants, the installed power of generator sets is chosen much higher than the maximum load power, which is associated with such a parameter for the units as Load Acceptance. For diesel units this parameter is 50-60% of the rated power, overestimating of power is justified when the load is variable. The utilization factor of the installed power of generator sets is often at the level of 20-25% in this case. The variable load is characteristic for a number of objects: mining, oil and gas, metallurgical, processing plants, etc. To ensure the stable operation of these objects, a large number of generating sets are often selected, which are constantly switched on to be able to provide a load altering by redistributing it between the generators. This operation mode is non-optimal in terms of fuel consumption and the cost of electricity. In addition, this mode intensively reduces the motor resource of the units, which also leads to high operating costs.

The use of gas is for energy supply of such objects a more attractive way of generating electricity in terms of minimizing fuel costs. But unlike diesel fuel, gas is more difficult to transport. A constant source is needed.

It is possible to re-equip part of the object to natural or associated gas, which is due to their small remoteness from natural deposits. In this case, the construction of gas mains from Siberia to the Far East and China («the Power of Siberia») can contribute to the usage of gas for power generation.

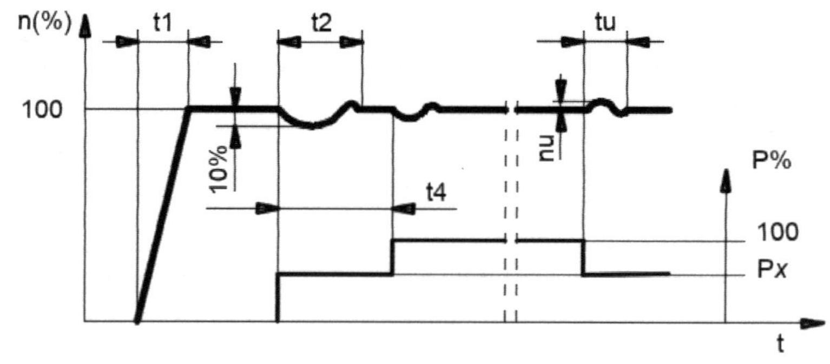

Figure 2. Transient parameters gas generator sets

Gas engine units have a much worse parameter of Load Acceptance, which is associated with greater inertia of the fuel path (Figure 2). Such units are less likely to perceive the outbursts and dumps of the load power, if the values shown in Table 1 (based on the example of a particular manufacturer) are exceeded, they are disconnected by the action of technological protections. The disconnection of one unit entails a chain reaction, leading to the shutdown of the whole power grid. Also, the gas engine units are basically designed in this way, [2] that the normal mode of operation is their loading in the power range from 40 to 100%.

Table 1. Transient parameters of gas generator sets.

Parameters	Symbol	Value
Frequency and voltage transient time during system start-up	t_1	24 s
Transient time when applying the load stage	t_2	10…12 s
Minimum time between load stages	t_4	15 s
Level of the load stage P_x (% of nominal)	P_1	30 %
	P_2	55 %
	P_3	75 %
	P_4	85 %
	P_5	Steps from 5 % to 100 %
Frequency overshoot at load shedding	nu	Less than 6.5 %
Transient time at load shedding	tu	0…13 s

3. Proposed Method

Compensation of power deviations by applying electrical energy storage system, will allow the gas engine power plant to operate in the zone of permissible parameters (Table 1). However, it is necessary to determine the amount of compensation. The complete compensation of the load variation on the one hand improves the operation of the generator sets, allowing to save fuel, and on the other hand leads to an increase in the energy capacity of the electrical energy storage system. Due to the high cost of batteries, this method is economically inefficient. Therefore, there is a need to minimize capacity by using special control algorithms.

We proposed the method which can be used to compensate of the rate of increase in power. The optimal value of optimal speed is set at the control system of EESS which determine the optimum changing of power of the generator set. If the load increases with a higher speed, EESS compensates the difference:

$$\frac{dP_{gen}}{dt} = \frac{d(P(t)_{load} - P(t)_{EESS})}{dt} \tag{1}$$

The proposed mode of operation allows the gas engine generator sets operate in an acceptable mode, while minimizing the energy capacity of electrical energy storage system. In this case, the required energy capacity of the electrical energy storage system can be calculated as:

$$E = \int_{t1}^{t2} (P(t)_{load} - P(t)_{gen})dt \tag{2}$$

The figure 3 illustrates the operation of EESS in the mode of compensation of power increase.

4. Experimental Validation

The proposed method of control was tested in an experimental setup including a 125 kVA diesel generator set, load modules simulating load oscillations, as well as an electrical energy storage system of 100 kVA and 153 kWh. Figures 3-4 show the experimental results of effect of the EESS during the reducing or increasing power load respectively. Figure 5 represents the single line diagram of experiment.

According to graphs the generating set in such modes is loaded smoothly, which confirms the effectiveness of the control by the proposed method. In the course of the experiment, fuel consumption measurements were also carried out, so the use of EESS allowed to reduce the fuel consumption by 5%.

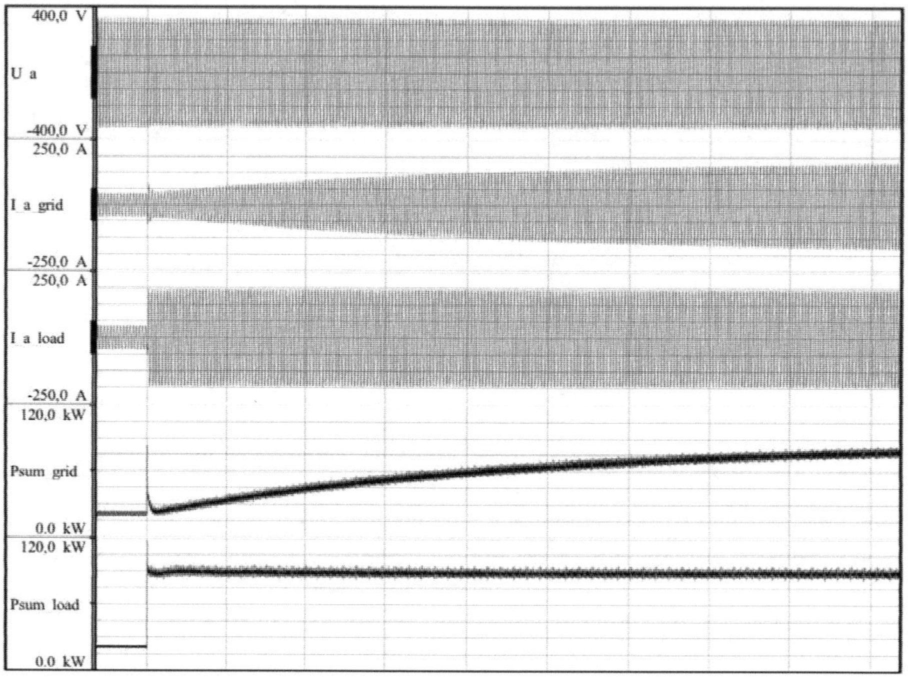

Figure 3. Response to the power draw at the generator and in conjunction with the EESS in the dP/dt compensation mode (from the top: the phase voltage diagram; phase current of the generator; the phase current of the load; the power of the generator; the load power)

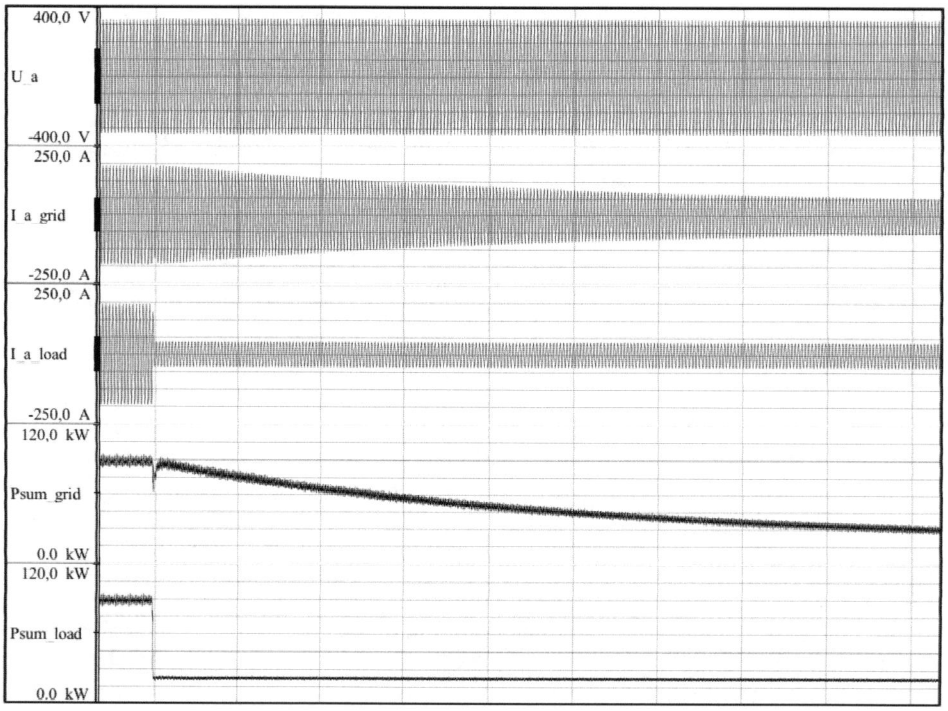

Figure 4. Response to the power reset at the generator and in conjunction with the EESS in the dP/dt compensation mode (from the top: the phase voltage diagram; phase current of the generator; the phase current of the load; the power of the generator; the load power)

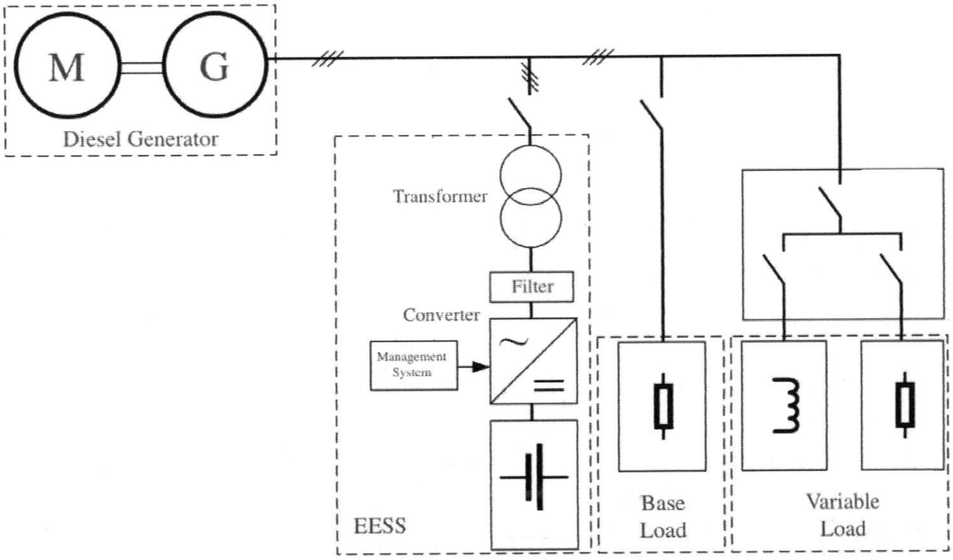

Figure 5. Single line diagrams of experiment

5. Conclusion

Electrical energy storage systems are multifunctional devices, the applications of which is extensive. The method proposed in this paper is effective for ensuring the stable operation of isolated power systems. However, due to the fact that the load oscillations are compensated, as much as is necessary for stable operation of the generators, the energy storage capacity of the electrical energy storage system can be chosen much less. Considering the fact that the cost of storage batteries is at a rather high level, the proposed method will reduce capital costs in the construction of a hybrid isolated power system.

In addition, positive effect is the possibility of replacing diesel generator sets with gas engines at objects where it was previously impossible in terms of operating modes of the power plant. Such solution will allow more than 2 times the fuel transportation costs to be considering, as well as reduce the emissions of toxic substances into the atmosphere, since natural gas is a more environmentally friendly resource compared to diesel fuel.

6. References

[1] Key World Energy Statistics, International Energy Agency, 2017, p.6

[2] Technical information from Cummins Power Generation Transient load capabilities of natural gas generator sets By Timothy A. Loehlein, *Technical Advisor in Sales Application Engineering*

[3] Brian Jabeck Market Development and Design Engineer Consultant Electric Power Division, *The Impact of Generator Set Underloading*, October 2014

[4] Application of a Resilience Framework to Military Installations: A Methodology for Energy Resilience Business Case Decisions N. Judson A.L. Pina E.V. Dydek S.B. Van Broekhoven A.S. Castillo/ *Technical Report Lincoln Laboratory*. Massachusetts Institute of Technology, Lexington, Massachusetts. 4 October 2016. p. 29

[5] Zobaa, A.F. Energy storage technologies and applications. Rijeka, Croatia: *Intech*, 2013., - 328 pp.

[6] Akagi. H. Instantaneous Power Theory and Applications to Power Conditioning / H. Akagi, E.H. Watanabe, M. Aredes // IEE Press, *John Wiley and Sons Inc.* 2007. – P. 389.

[7] Dybko M. A. Active power filter with battery energy storage based on NPC inverters / M. A. Dybko, S. V. Brovanov // *16 International conference of young specialists on micro/nanotechnologies and electron devices (EDM)*: [proc.], Altai, Erlagol, 29 June – 3 July 2015. – IEEE, 2015. – P. 415-421.

IOP Conf. Series: Materials Science and Engineering **836** (2020) 012008 doi:10.1088/1757-899X/836/1/012008

Identification of parameters for the hybrid electrical energy storage system in autonomous power system

S.A. Eroshenko[1], G. B. Nesterenko[2,3], V. M. Zyryanov[2], N. G. Kiryanova[2], G. A. Prankevich[2,3]

[1] Ural Federal University, 620002, Ekaterinburg, Russian Federation
[2] Novosibirsk State Technical University, Novosibirsk, 630073, Russian Federation
[3] Energy Storage Systems LLC, Novosibirsk, 630007, Russian Federation

nesterenkogb@yandex.ru

Abstract. The article describes the approach to the selection of parameters of a hybrid electrical energy storage system (EESS) based on supercapacitors (SC) and lithium ion battery (LIB) for an autonomous power station with a diesel generator set (DGS). The calculations of transients in an autonomous power supply system with an abruptly variable load and participation of a hybrid EESS. The procedure of determining the share of participation in the energy exchange of storage subsystems based on LIB and SC is presented.

1. Introduction

Currently, electric energy storage systems (EESS) are widely used to solve various problems of the electric power industry. In recent decades, the intensive development of energy storage technologies has led to the creation of EESS with characteristics (power, energy intensity, efficiency factor, speed) that allow implementing projects with technical and economic efficiency. In 2017, the Ministry of Energy of the Russian Federation approved the Concept for the Development of the Market for Electricity Storage Systems in the Russian Federation [1]. In addition, specific tasks for the introduction of energy storage systems in the Russian Federation energy sector are indicated in the Energy Plan, which is part of a long-term comprehensive program of the National Technology Initiative that aimed the formation of fundamentally new markets and the creation of conditions for Russia's global technological leadership by 2035 [2].

Modern fast-speed EESS are fundamentally new energy power devices designed for controlled energy exchange with the power system in order to organize the desired mode or to control dynamic processes. EESS is able to almost instantly control the active power balance according to any given algorithm. In accordance with the given task, EESS can be used as a device for reactive power compensation, active filter of higher harmonics and as a means of compensating for asymmetries in three-phase networks.

Due to the novelty of the EESS technologies, their development and implementation in the practice of the Russian electric power industry begins with relatively small rated power and energy intensity. There are number of projects with EESS in autonomous energy systems of Russian Federation for implementation, which are characterized by high economic and technical efficiency.

The speed of power change in the process of controlled energy exchange is determined by the functional purpose of EESS. The most relevant types of energy storage devices currently are: lithium-ion batteries and supercapacitors. The first type is most effective for relatively slow processes, and the second for faster. If it is necessary to control both fast and slow processes, it is advisable to use a hybrid EESS, which includes both types of energy storage devices.

The combination of various energy storage technologies in hybrid systems allows to create more advanced EESS to solve a wide range of problems. Hybrid EESS combine technologies that can complement and mutually compensate for the negative characteristics of each other's characteristics [3]. In the case of application of the storage subsystem based on SC and LIB in a hybrid EESS: a high SC resource in terms of the number of work cycles and lower LIB cost are successfully combined. At present, hybrid EESS are used in the microgrids field in combination with renewable energy sources [4], in the operation of electric vehicles and as part of charging stations [5].

The choice of the parameters of each storage elements in the energy storage subsystem is one of the main issues to be solved in the process of design a hybrid EESS. The article outlines an approach to the selection of parameters of batteries and supercapacitors for hybrid EESS, based on the analysis of load diagrams.

2. Rated power, available power and energy intensity

The two main parameters of any type of EESS are rated power and energy intensity. The first parameter determines the power value for which all power elements of EESS are designed. The second parameter requires further explanation.

Firstly, due to to technological limitations it is impossible to use the full energy intensity in energy exchange in the process of charge and discharge.

Secondly, the life of accumulating elements depends on the depth of the discharge during the charge/discharge process. SC can withstand a large number of complete charge/discharge cycles (at least 500,000). The number of LIA cycles at a discharge depth of 80% is 3000–4000, after which the battery structure is irreversibly degrading [6].

For SC the quantity of exchange energy depends on the operating voltage range on the DC side of the reversing converter. If the ratio of the minimum and maximum operating voltage is assumed to be 0.6 [7], EESS can be used for exchange with the power system 64 % of the nominal energy consumption of the SC module (taking into account the nonlinearity of the dependence of the stored energy in the SC on voltage). The charge level should not be less than 10% and more than 90% in normal operation of LIB [8].

The degree of charge of the accumulation subsystem is constantly changing in the process of energy exchange. It leads to a change in the maximum value of power that is able to consume / give off the EESS in the charge/discharge mode. This power value is available power. The power of EESS can take any value in the range from 0 to the value of the available power in the process of control. And the available power continuously changes with a change in the degree of charge.

3. Parameters choice for hybrid EESS

It is necessary to select and coordinate the power and energy consumption of both energy storage subsystems for calculating the parameters of the hybrid EESS. In addition, it is necessary to solve the question of the distribution of participation volumes between the storage subsystems in the process of management of energy exchange of EESS with the power system. One of the methods for distributing energy exchange power in a hybrid EESS is linear filtering [9]. In this case, control actions are formed using filters that distribute power between storage subsystems depending on their response time. Low-frequency disturbances are assigned to LIB, disturbances of a higher frequency are assigned to SC.

The calculation of the parameters for the hybrid EESS is shown on the example of an autonomous power system of floating crane with a diesel generator set (DGS) with a rated power of 300 kW. A feature of the crane operation is an abruptly variable load schedule with a low installed capacity utilization factor of DGS. Figure 1 shows that the DGS power varies over a wide range, while its average value is 66.7 kW. The abruptly variable nature of the load leads to an increase in diesel fuel consumption and a decrease in diesel engine life [10].

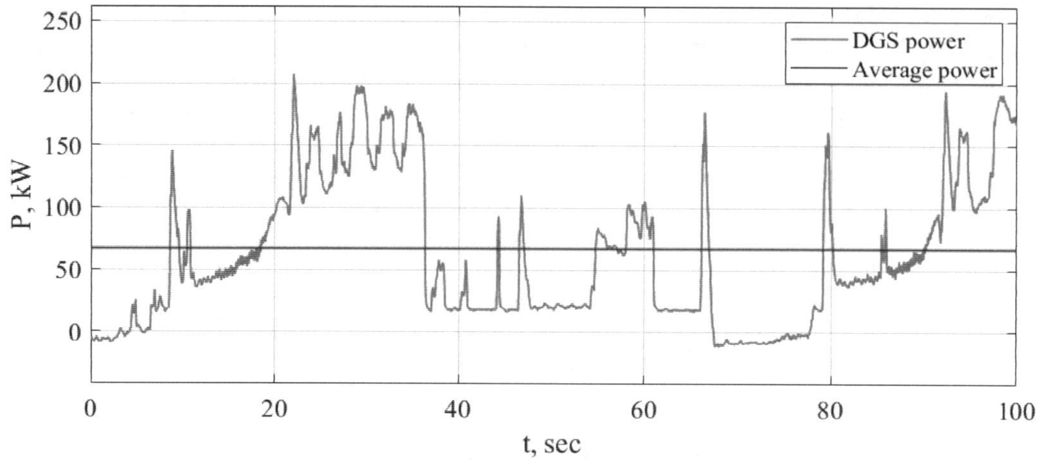

Figure 1. Fragment of a load schedule of a DGS.

Combined use of EESS and DGS provides replacement 300 kW of DGS rated power with a 100 kW DGU, while a significant part of the variable component of the load diagram must be compensated by the storage system. Due to the fact that the load diagram contains a wide range of frequencies, it is advisable to use a hybrid EESS. Compensation of the low-frequency part of diagram spectrum and limitation of the DGS power deviations from a given average value is assigned to LIB, and compensation of the high-frequency part is assigned to SC. This circumstance is due to necessity of greater energy intensity for limiting power and aligning the main part of the load schedule compared with compensation for power fluctuations of a higher frequency. In addition, the cost of SC is significantly higher than LIB.

To implement the above, the following operating conditions are specified for the LIB-based storage subsystem: when the load power is exceeded 86.6 kW or less than 46.6 kW, the LIB-based storage subsystem compensates for the difference between the load power and the indicated power limitations. The power of DGS should not go beyond the corridor limited by these power values. It is proposed to entrust the SC with the task of damping the frequencies of power fluctuations in the range of 0.4-2 Hz, which coincide with the natural frequency of oscillation of the generator unit [11].

According to [12], the frequencies of periodic changes in power above 2 Hz do not significantly affect the generating units due to their inertia, and the speed controllers of the generating units quite effectively cope with frequencies below 0.4 Hz.

The parameters for the hybrid EESS of each of the energy storage subsystems are selected according to the method described in [13]. The following parameters were selected: for LIB power – 150 kW, exchange energy intensity – 0.4 kW·h, for SC power – 40 kW, exchange energy intensity – 0.06 kW·h. The exchange energy intensity is understood as the minimum amount of energy that EESS must exchange with the energy system in the process of power control.

3.1. Calculation experiment

The calculations of the operation of an autonomous power plant of a crane together with a hybrid EESS are performed with the using of the mathematical model of the EESS developed by the authors [14]. Figure 2 shows the load power and the power of DGS, and also shows the average value of the load power. As follows from the figure, the hybrid EESS quite satisfactorily completed the task. The participation of each storage subsystem of the EESS for capacity management is shown in figure 3.

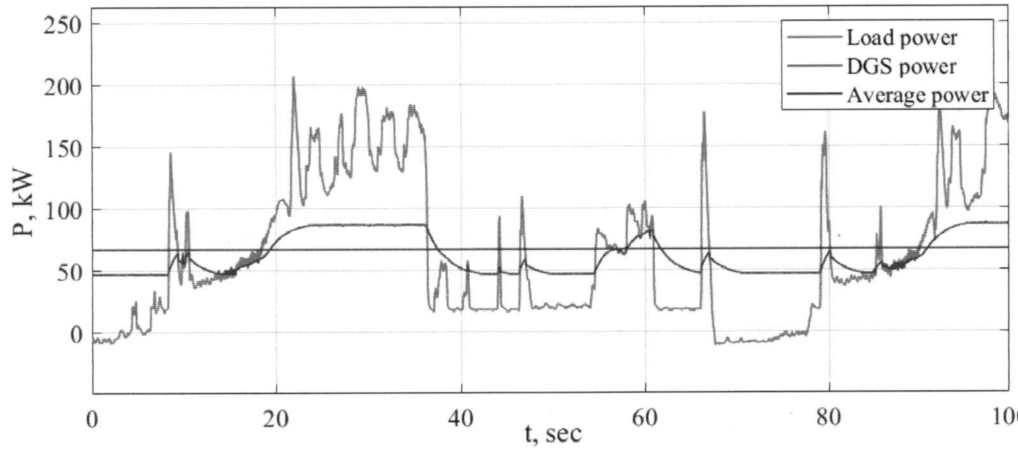

Figure 2. Calculation results for load and DGS power with participation EESS.

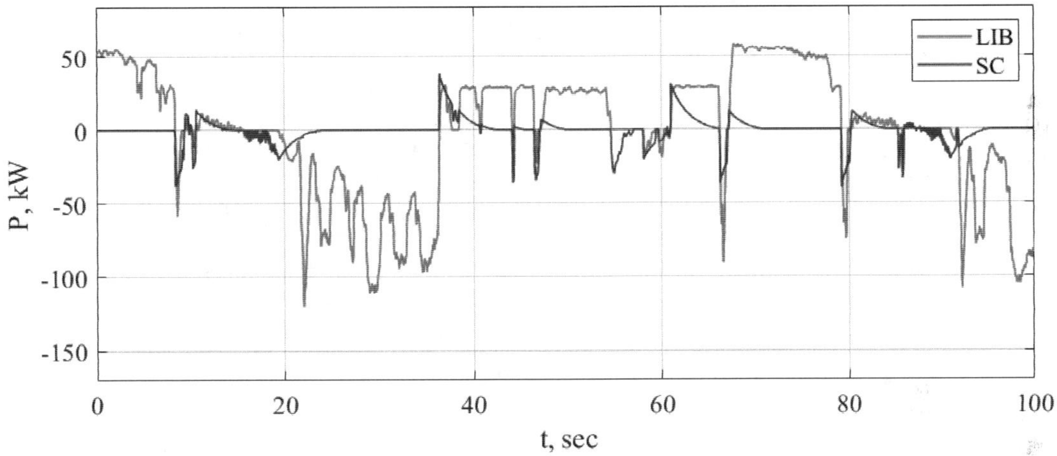

Figure 3. Power of the storage subsystems of hybrid EESS.

The calculation results show a high efficiency of power control using a hybrid EESS. The main part in the management in this example is taken by the LIB-based accumulation subsystem. The subsystem based on SC accounts for the development of sharp changes in power with large values of *dP/dt*.

In this example, when choosing the parameters of the subsystems of accumulation, economic issues were not taken into account. At the same time, this task becomes a technical and economic one taking into account the significant difference in the cost of LIB and SC modules. It is required to take into account the permissible discharge depth and the service life of each of the subsystems. And since this is connected with the choice of energy intensity, the cost of a hybrid EESS should be optimized, provided that a given technical effect is obtained.

4. Conclusion

The main task of choosing the parameters of a hybrid EESS is to agree on the share of each of the subsystems of energy storage in the overall energy exchange of EESS with the power system. It is necessary to take into account the resource of the accumulating elements of each accumulation subsystem by the number of charge / discharge cycles, the permissible discharge depth, the number and arrangement of elements, their cost. Thus, such task is techno-economic.

The variant of calculating the parameters of the hybrid EESS presented in the article describes the technical side of the issue. The procedure for selecting parameters and coordinating the volumes of participation of various types of storage subsystems in the energy exchange process shown using a concrete example of autonomous power system. The considered version of EESS combines the advantages of both types of storage subsystems: the performance of supercapacitors and their high resource in terms of the number of duty cycles, on the one hand, and greater energy consumption at a lower cost of lithium-ion batteries, on the other hand.

References

[1] The concept of development of the market for electric storage systems in the Russian Federation 2017 *Ministry of Energy of the Russian Federation* p 43

[2] Action Plan ("roadmap") of the Energy of the National Technology Initiative 2016 *Approved by the Presidium of the Presidential Council on Economic Modernization and Innovative Development of Russia Protocol No. 4*

[3] Jérémy Dulout, Bruno Jammes, Lionel Séguier, Corinne Alonso 2015 Control and design of a hybrid energy storage system *17th European Conference on Power Electronics and Applications (EPE'15 ECCE-Europe) (Geneve)* DOI: 10.1109/EPE.2015.7309468

[4] Jianwei Li, Bertrand Cornelusse, Philippe Vanderbemden, Damien Ernst 2017 A SC/battery Hybrid Energy Storage System in the Microgrid *Energy Procedia* vol 142 pp 3697-3702 DOI: 10.1016/j.egypro.2017.12.264

[5] Ziyou Songa, Jianqiu Lib, Jun Hou, Heath Hofmann, Minggao Ouyang, Jiuyu Du 2018 The Battery-Supercapacitor Hybrid Energy Storage System in Electric Vehicle Applications: A Case Study *Energy* vol 154 pp 433-441 DOI: 10.1016/j.energy.2018.04.148

[6] A. Voroshilov, A. Petrov, E. Chudinov 2017 Lithium iron phosphate battery. Modeling the charging mode *News of Electrical Engineering* No. 2(104)–3(105) pp 44-49

[7] Bidirectional voltage converter manufactured: catalog 2018 *LLC "DC system" (SPT)* p 2

[8] GOST R 58092.1–2018 *ELECTRIC ENERGY STORAGE SYSTEMS (SNEE). Terms and Definitions* (M .: Standartinform) p 54

[9] V.G. Chirkin, L.Y. Lezhnev, D.A. Petrichenko, I.A. Papkin 2018 A Battery-Supercapacitor Hybrid Energy Storage System Design and Power Management *International Journal of Pure and Applied Mathematics* vol 119 No. 15 pp 2621-2625

[10] V.A. Alemasov, D.I. Baranov, M.D. Bimukhanov, V.M. Zyryanov, A.B. Mosienko 2007 The effectiveness of energy storage in autonomous power systems with a rapidly changing load *Scientific problems of transport in Siberia and the Far East* No. 2 pp 113-120

[11] D.Yu. Baluev, V.M. Zyryanov, N.G. Kiryanova, G.A. Prankevich 2017 The use of energy storage for damping power fluctuations in autonomous energy systems *Infrastructure sectors of the economy: problems and development prospects: collection of articles* (Novosibirsk) pp 181-187

[15] N.G. Kiryanova, D.Y. Baluev, G.A. Prankevich, V.M. Zyryanov 2017 Energy storage device application for load oscillations damping in isolated power systems *Actual issues of mechanical engineering (AIME 2017) : proc. of the intern. conf.*, (*Advances in Engineering Research* vol. 133) pp 325-330 DOI: doi:10.2991/aime-17.2017.53

[16] V. M. Zyryanov, N. G. Kiryanova, G. A. Prankevich, D. Yu. Baluev 2018 The method of selecting the parameters of the energy storage devices with an abrupt variable load *Bulletin of Irkutsk State Technical University* No. 5 pp 105-114 DOI: 10.17212/1814-1196-2018-2-131-142

[17] Bachurin P.A., Zyryanov V.M., Kiryanova N.G., Kuchak S.V., Metalnikov D.G., Nesterenko G.B., Potapenko A.M., Prankevich G.A. 2018 Mathematical model of the energy storage system in the power system *Actual problems of electronic instrument engineering (APEIE – 2018) (Novosibirsk) pp.* 228 – 234 DOI: 10.1109/APEIE.2018.8545298

IOP Conf. Series: Materials Science and Engineering **836** (2020) 012009 doi:10.1088/1757-899X/836/1/012009

Vibration problem solution of high-capacity hydropower units

A.I. Khalyasmaa[1], S.M. Sultonov[2], J.S. Ahyoev[2], A.H. Bahovadinov[3]

[1]Ural Federal University, Russian Federation, Ekaterinburg
[2]Tajik Technical University, Department of Power Stations, Dushanbe, Tajikistan
[3]Sangtuda-2 HPP, Engineering and Maintenance Department, Tajikistan

a.i.khaliasmaa@urfu.ru

Abstract. This paper proposes hydro generator vibration observations under various conditions and solution of a problem of high vibration in upper guide bearing of 110MW hydro generator. Among the test, which were carried out to identify vibration origin of hydro generating unit were: voltage drop tests, magnetic flux measurements, interferric gap measurements, turbine pressure pulsation measurements, etc. It was shown, that among other possible reasons, correct balancing, including pivot bearing adjustment was the most efficient measure to eliminate vibration in start-up modes of hydro power unit. According to the results of conducted step-by-step defect analysis, the corresponding recommendations were provided for service and maintenance regarding the generation unit No2 of Sangtuda-2 Hydro Power Plant, Tajikistan.

1. Introduction

Vibration monitoring system is the most powerful device for state tracking of rotating machines during their operation. Due to online monitoring of vibration and on basis of programmed routine attention, devastating failures of machine will be prevented.

Sangtuda-2 HPP unit's vibration system monitors relative and absolute vibration at three points: turbine guide bearing, thrust bearing and upper guide bearing.

To measure the absolute vibration in bearings, a displacement type of sensor is used due to the lowness of unit's speed (71.4 rpm). The type and characteristic of sensors represented below:

- Relative (run out) sensors:
  - measurement principle: eddy current;
  - frequency range: 0 – 10kHz(-3dB);
  - measuring range: 2mm;
  - sensitivity: -8mV/um(-200mV/mil).
- Absolute vibration sensor:
  - measurement principle: displacement;
  - frequency range: 0.5 – 150Hz(-3dB);
  - measuring range: ±1000um;
  - Sensitivity: 8V/mm±5%.

2. Identification of the problem

During vibration monitoring of unit 2, vibration in pilot bearing raised and shut up to the alarm level, following which at some conditions an alarm signal was received. Vibration alarm was recorded by relative sensors, especially at Y-axes. Figure-1 illustrates condition of unit during startup, rolling and

IOP Conf. Series: Materials Science and Engineering **836** (2020) 012009 doi:10.1088/1757-899X/836/1/012009

standby modes. During the unit startup, relative vibration of X, Y axes and absolute vertical vibration raised, but horizontal vibrations (X and Y axes) maintain normal. This condition is the effect of unit's run, so when the unit reaches the rated speed (rolling mode fulfilled), the state of the unit becomes normal. After excitation system initiates (standby mode fulfilled), all conditions of vibration become worse.

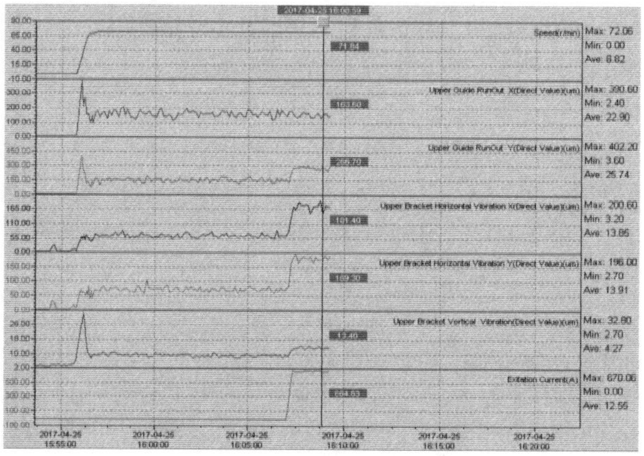

Figure 1. Vibration characteristic at upper guide bearing.

2.1. Voltage drop test of rotor for ensuring equal impedances of poles

Vibration sensors check of their normal operation together with correct earthing of screen cable is higher at standby condition than during rolling. The main causes of higher vibration at standby mode can be disbalanced magnetic field in salient pole rotors. One of the causes of disbalanced magnetic field is short circuit in rotor's poles. To identify short circuits in rotor's poles, a rotor voltage drop test is used, that measures each pole's impedance. Test results showed in Table-1. They depict that impedances of poles are equal and are not short circuited in their standstill condition.

2.2. Measurement of magnetic flux to ensure accurateness of poles during rotating

Although, voltage drop test is not comprehensive because of the rotor's rotation and the possibility of occurrence of short circuits only during its running due to the action of the centrifugal force on the rotor. For ensuring accuracy of poles during rotor's rotating, magnetic flux of poles is being monitored [1,2]. To measure the magnetic flux of poles, flow transducer is glued on stator core. Output of probe is being monitored and registered by capacity. Figures 2 and 3 illustrate induced voltage in flow transducer and calculated magnetic flux density at a holding mode of generator. Each of these graphs shows the symmetry values between pair poles that demonstrate the accuracy of poles.

Table 1. Voltage drop test with 2A ac current.

I=2,00 A V_{total} = 124,4 V								
Pole No.	Voltage (V)	Impedance (Ohm)	**Pole No.**	Voltage (V)	Impedance (Ohm)	**Pole No.**	Voltage (V)	Impedance (Ohm)
B1 – B82	1,432 -1,500	0,716 -0,750	B29 – B56	1,455 -1,512	0,728 -0,756	B57-B84	1,417 -1,529	0,709 -0,765

2.3. Interferric gap measurement

Another cause of disbalanced magnetic field can be disharmony of the interferric gap between stator and rotor [3]. The result of measurement was compared with commissioning protocol that showed normal condition of the air gap between stator and rotor.

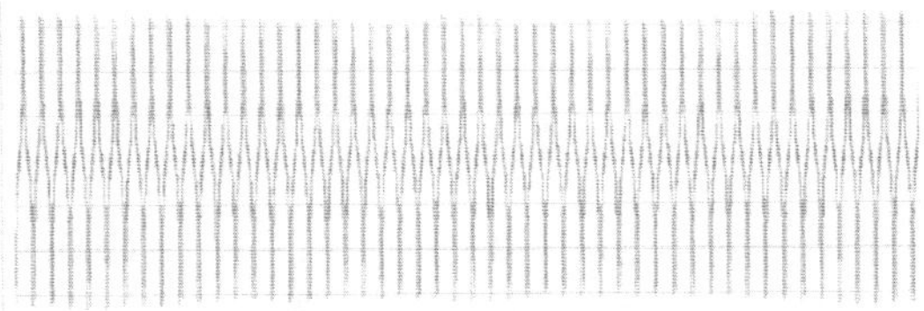

Figure 2. Induced voltage in probe at standby mode of generator.

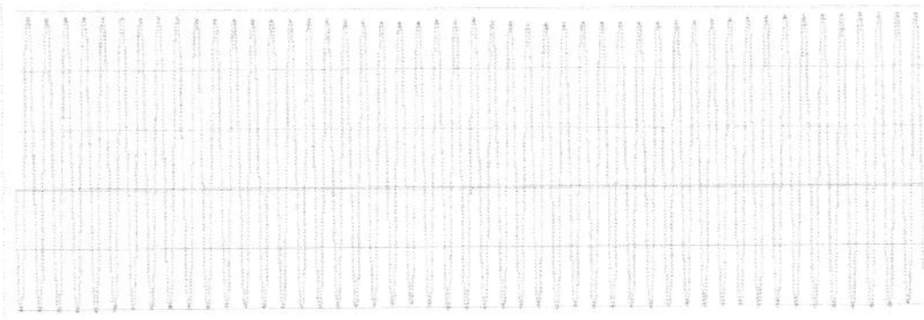

Figure 3. Magnetic flux density calculated from induced voltage at standby mode of generator

2.4. Turbine pressure pulsation measurement

Study of rotor poles and interferric gap between stator and rotor showed that the state of interferric gap or rotor poles have not caused increasing of vibrations during the years, also vibration of the unit, especially upper guide bearing's vibration changed according to different conditions of power outturn and net head. After ensuring the operational capability of rotor poles and interferric gap, pressure pulsation data of turbine together with net head and blade angle was checked. Via this test, table of correlation between blade angle during the opening of wicket gates and net head that were used in turbine's operation system, was checked with optimum points. The optimum point for blade angle regarding to opening wicket gates and net head is the point in which correlation of vibration of the unit and pressure pulsation in turbine is optimum. This test showed that the table, used in system of regulation, does not differ from optimum point which leads to high vibrations.

2.5. Balance weight installation in rotor

The next step to improve the vibration condition in upper guide bearing was to install the balance weight. After studying of FFT data of vibration in upper guide bearing by engineering department, balancing weight was installed in field spider. The result of installation of balancing weight is illustrated in Figure-4. These graphs show better condition than previously, but vibration is still higher than normal and sometimes depending on different conditions alarm level of relative vibration (run out) Y-axis is being received.

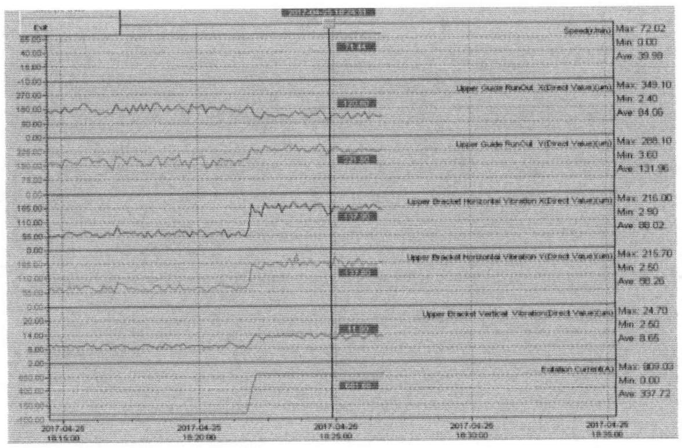

Figure 4. Magnetic flux density calculated from induced voltage at standby mode of generator.

2.6. Adjustment of pilot bearing

Secondly, after checking the vibration sensors, the adjustment of pilot bearing was made. Clearance between upper bearing's bush was checked by setting into operation. There were no differences that would lead to worse vibration conditions.

During the next annual routine attention of unit-2 (at 2018) pilot bearing was readjusted. Shoe clearance was adjusted to 0.17 mm. Due to this action vibration of the unit was brought to a better condition, but Y-axes relative vibration stayed high and close to the alarm level. This situation led to an overall check of mechanical joints of generators. Tightness of all bolts, nuts and latches of generator was checked.

During the check of upper bracket that has a damper in the end of each arm (Figures 5 and 6 illustrate upper bracket and jack), one of these joints was not adjusted and did not work normally.

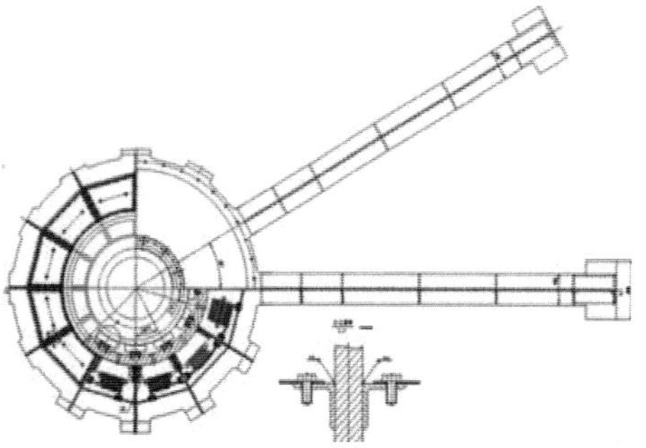

Figure 5. Upper Guide bearing with bracket.

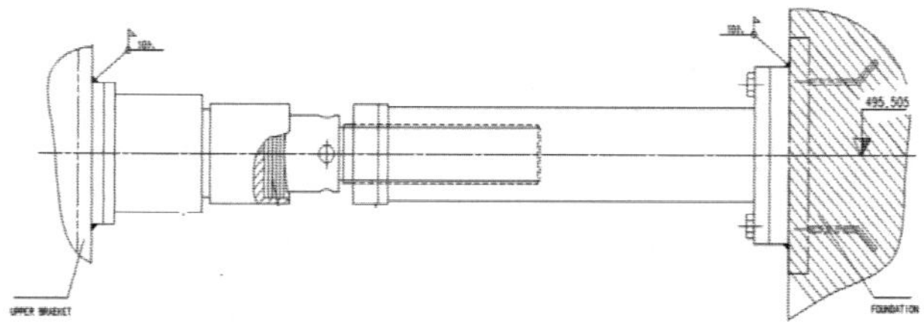

Figure 6. Upper bracket dumper jack.

After damper jack was stripped down, then adjusted and put back in its position, unit is being started together with vibration check. The result of vibration records is illustrated in Figures 7 and 8. The graphs show that vibration value and disturbances decreased.

The value of vibration in Y-axis before adjustment of jack was about 250um when X-axes vibration was about 160um. After installation, the value of vibration in both axes is about 150-160um at standby mode. Vibration of the unit in other conditions (different powers and net heads) was also checked and problems were not detected.

3. Conclusion

Accurate, correct and, of course, comprehensive maintenance leads to reliable work of equipment. Together with comprehensive routine attention, reliable work of equipment can be guaranteed.

On-line monitoring of vibration is a powerful device for detecting great majority of injuries of rotating machine.

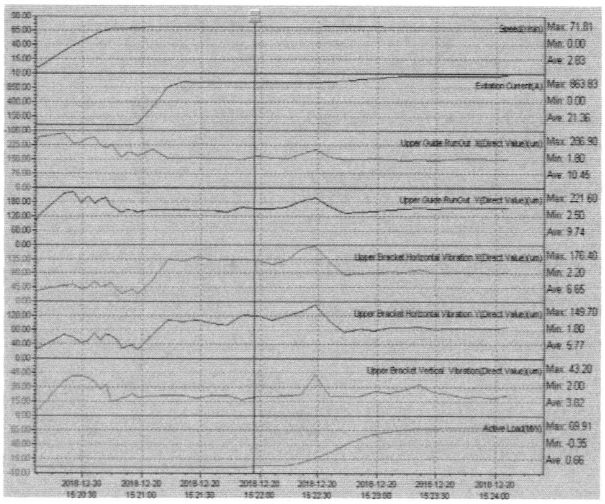

Figure 7. Vibration graph after jack adjustment during start-up and loading.

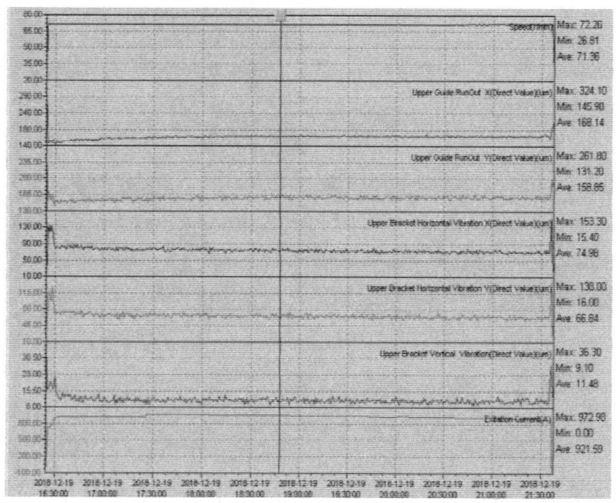

Figure 8. Vibration graph after jack adjustment during start-up and 70 MW load.

References

[1] G. C. Stone, M. Sasic, J. Stein and C. Stinson, "Using magnetic flux monitoring to detect synchronous machine rotor winding shorts," Conference Record of 2012 Annual IEEE Pulp and Paper Industry Technical Conference (PPIC), Portland, OR, 2012, pp. 1-7.

[2] Steve R. Campbell, Greg C. Stone, Evens Jourdain,Jan Stein, "Initial Experience With On-Line Hydrogenerator Rotor Winding Condition Assessment UsingFlux Monitoring" 2006 Doble Engineering Company-73rd Annual International Doble Conference, All Rights Reserved.

[3] Blagoje Babic, Nenad Kartalovic, Savo Marinkovic, Dejan Misovic, Dragan Teslic, Zorica Milosavljevic, Aleksandar Nikolic, "Correlations between Magnetic and Vibration Measurements on Hydro generators", Recent Advances in Intelligent Control, Modelling and Simulation, ISBN: 978-960-474-365-0, pp. 171-175.

[4] Griscenko M. Air Gap Monitoring of Hydropower Unit Generator to Advance Vibration Diagnostic Procedure. Summary of the Doctoral Thesis. — R.: RTU, 2015. — 20 p.

[5] Geraldo Carvalho Brito Junior, Roberto Dalledone Machado, Anselmo Chaves Neto, Mateus Feiertag Martini, "Experimental Aspects in the Vibration-Based Condition Monitoring of Large Hydro generator s", Hindawi, International Journal of Rotating Machinery Volume 2017.

[6] Rati Kanta Mohanta, ThangaRaj Chelliahm, Srikanth Allamsetty, Aparna Akula, Ripul Ghosh, "Sources of vibration and their treatment in hydropowerstations." Engineering Science and Technology an International Journal 20 (2017) 637–648 pp.

[7] Azuaje C. and Millan A., "Stator Deformation of Large Hydrogenerators and Its Effects on the Machines", Transmission & Distribution Conference and Exposition: IEEE/PES, pp. 1-5.

[8] Mihaela Răduca, Eugen Răduca, Iancu Tătucu, "The checking at vibrations of hydro generators stator steel parts", The 4th Symposium with International Participation, Palic, Serbia & Montenegro, vol. 4, pp. 225-228.

[9] Jules Pascal, "Vibrations from Magnetic Forces in Hydropower Generators", Norwegian University of Science and Technology, Department of Electric Power Engineering, Master of Science in Electric Power Engineering.

[10] Nowicki, Ryszard & Raegan Macvaugh, "XY Measurements for Radial Position and Dynamic Motion in Hydro Turbine Generators," ORBIT, Volume 30 Number 1, 2010.

Digital twin process and simulation operation control technology for intelligent manufacturing unit

Yichao He[1], Niansong Zhang[2] and Aimin Wang[3]

[1]Department of Mechanical Engineering, Nanjing University of Science and Technology, No.200 Xiao Lingwei Street, Xuan Wu District, Nan Jing City, Jiang Su Province, China
[2]Department of Mechanical Engineering, Nanjing University of Science and Technology, No.200 Xiao Lingwei Street, Xuan Wu District, Nan Jing City, Jiang Su Province, China
[3]Department of Mechanical Engineering, Beijing Institute of Technology, 5 South Zhongguancun Street, Haidian District, Beijing 100081, China

Email: yichaoh@163.com

Abstract. In the manufacturing process, the intelligent manufacturing unit will produce unexpected situations with different forms and a large amount of data. How to control the intelligent manufacturing unit more intuitively and effectively control the operation and control of the trolley is an urgent problem to be solved. This paper proposes a digital twin process and simulation operation control technology for intelligent manufacturing units, including virtual workshop and actual workshop real-time mapping technology and digital twin-based simulation control technology. Through these two technical means, intelligent manufacturing unit realized visual real-time management and control. In the end, the author designed an intelligent manufacturing unit management system to successfully verify the effectiveness of the above technology.

1. Introduction

At present, there have been many studies about the application of digital twin technology domestic and abroad. The concept of digital twin was first proposed by Professor Grieves in the product lifecycle management course at the University of Michigan in 2003 and was defined as a three-dimensional model, including physical products, virtual products, and connections between them[1]. Domestic Cunbo Zhuang, Jianhua Liu, etc.[2] proposed the concept of digital twin technology, and pointed out that digital twins are technologies, processes and methods, and digital twins are objects, models and data. It is pointed out that digital twinning technology not only uses human existing theory and knowledge to build virtual models, but also uses virtual model simulation technology to explore and predict the unknown world, and constantly stimulates human innovative thinking, which provides new ideas and tools for the current manufacturing innovation. Fei Tao [3] proposed the implementation mode of the digital twin workshop, and defined its system composition, operation mechanism, characteristics and key technologies, which provided a theoretical and method reference for the realization of CPS in the manufacturing workshop.

Digital twin is a virtual model that creates physical entities in a digital way. It simulates the behavior of physical entities in real-world environments by means of data, and adds or expands physical entities through means of virtual and real interaction feedback, data fusion analysis, and decision iterative optimization, add or extend new capabilities for physical entities. As a technology that makes full use of models, data, intelligence and integrated multidisciplinary, digital twin provides a more real-time, efficient and intelligent service for the whole life cycle of products and the bridge and the link between

IOP Conf. Series: Materials Science and Engineering **836** (2020) 012010 doi:10.1088/1757-899X/836/1/012010

the physical world and the information world[3-5]. The first task of implementing a digital twin application is to create a digital twin model of the object. At present, the digital twin model mostly uses the three-dimensional model originally defined by Professor Grieves, namely physical entities, virtual entities and the connection between them [6]. However, most of the digital twin in practical applications mainly focus on three-dimensional visualization of the workshop and manufacturing information. For example, many of them have done three-dimensional visualization of the workshop, think that this is digital, but it is not. Digital twin emphasizes the synchronization between virtual and real, the operation of the actual workshop is consistent with the operation of the virtual workshop, and the virtual workshop can receive information from the actual workshop and make adjustments. Similarly, the actual workshop can make changes which are based on the commands issued by the virtual workshop.

This paper will focus on the problem of visual control of intelligent manufacturing unit of an enterprise, combined with the development of management software, focusing on the digital twin process and simulation operation control technology of intelligent manufacturing unit to realize the high-efficiency control of the intelligent manufacturing unit.

2. Digital twin process and simulation operation control technology framework

Through the SolidWorks software, the machine equipment, the staff and the AGV and other equipment facilities in the intelligent manufacturing unit are scaled and 3D modeled, and then saved into the step format and imported into the 3ds MAX software, and then exported after the software is rendered. Finally, it is imported into Unity software to build a virtual intelligent manufacturing unit. Finally, it is embedded into the control system of WinForm, and the data from the intelligent manufacturing unit is accepted through the control system, and feedback to the intelligent manufacturing unit in real time. The digital twin process and simulation operation control technology for intelligent manufacturing units mainly study how to realize the virtual and real synchronization between the virtual workshop and the actual workshop, and realize the data transmission and feedback. Therefore, the part of the virtual intelligent manufacturing unit is not excessively described. This paper will briefly explain the framework of digital twin process and simulation operation control technology in conjunction with Figure 1.

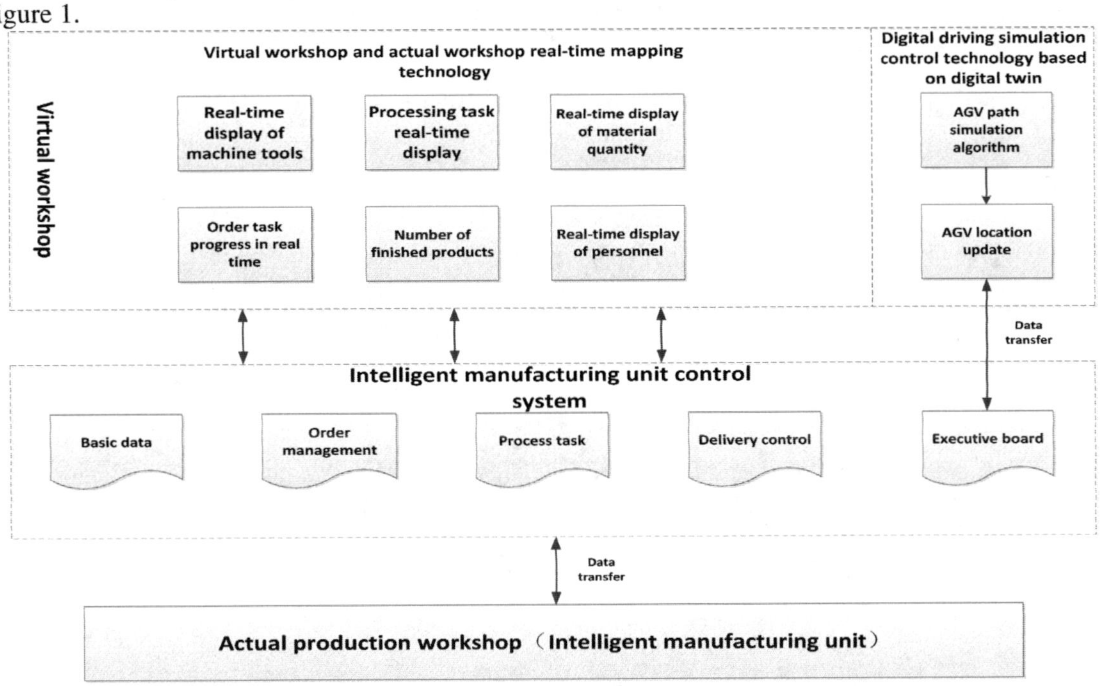

Figure 1. Digital Twin Process and Simulation Operation Control Technology Framework.

The technical framework of the digital twinning process and simulation run control technology for the intelligent manufacturing unit shown in Figure 1 is mainly composed of two technologies: virtual workshop and actual workshop real-time mapping technology and digital driving simulation control technology based on digital twin. The main applications are as follows:

1. Virtual workshop and actual workshop real-time mapping technology. Real-time mapping and virtual reality synchronization are two basic characteristics of digital twin. The core of the data is the data-driven model, which is to obtain real-time transformed data from the actual workshop, and transmit it to the virtual workshop in real time through the data transfer station, so as to realize real-time virtual workshop and actual workshop mapping technology.

2. Digital driving simulation control technology based on digital twin. The intelligent manufacturing unit contains the AGV. The AGV in the actual workshop runs synchronously with the 3D model of the AGV in the virtual workshop, but this is not enough. The fundamental purpose of digital twin is decision analysis, so it is extremely critical for the simulation operation control of the AGV. The material transported by the AGV can be simulated in the virtual workshop in advance, and the optimal solution is selected, thereby improving the working efficiency of the AGV and further improving the production efficiency of the intelligent manufacturing unit. In view of this, this paper proposes digital driving simulation control technology based on digital twin.

3. Key technology

The foregoing has briefly introduced the role of two key technologies in the overall framework of digital twin process and simulation operation control technology for intelligent manufacturing unit. The specific implementation principles of these two key technologies are described in detail below.

3.1. Virtual workshop and actual workshop real-time mapping technology

In order to realize the real-time synchronization between the virtual workshop and the actual workshop, it is convenient for the manager to view the real-time information of the workshop and the management workshop. This paper proposes the real-time mapping technology between the virtual workshop and the actual workshop. In addition to the three-dimensional modeling of all objects in the actual workshop, all the information in the workshop needs to be summarized and organized to clarify which data needs to be transmitted to the virtual workshop to implement the data-driven model. The specific implementation process of the virtual workshop and the actual workshop real-time mapping technology is shown in Figure 2.

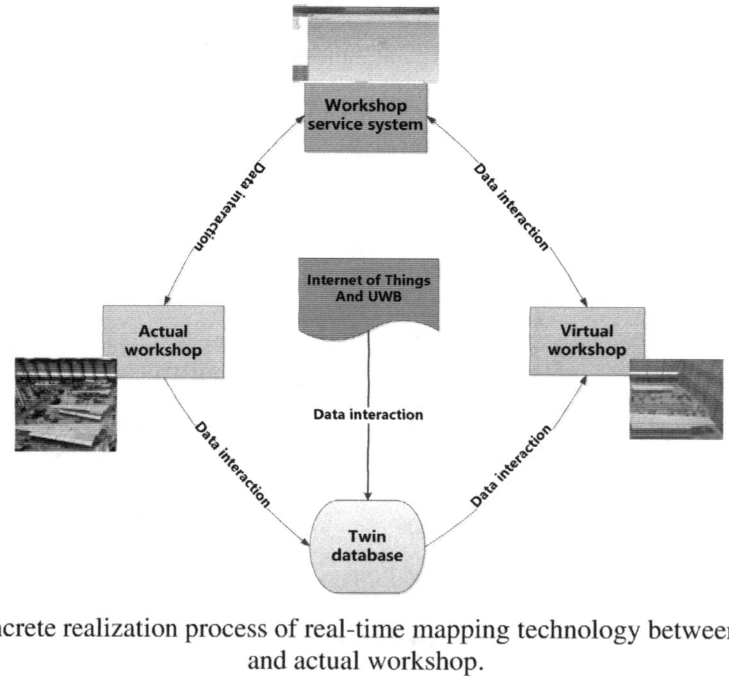

Figure 2. The concrete realization process of real-time mapping technology between virtual workshop and actual workshop.

IOP Conf. Series: Materials Science and Engineering **836** (2020) 012010 doi:10.1088/1757-899X/836/1/012010

In order to realize the synchronous operation of the virtual workshop and the actual workshop, the 3D models in the virtual workshop need to be driven according to the data of the real-time transformation of the actual workshop. According to the data source shown in Figure 2, there are two parts, namely the data of the workshop service system and the data in the twin database. The data in the twin database is mainly through the Internet of Things and UWB deployed in the actual workshop. The Internet of Things and UWB can monitor the movements in the actual workshop in real-time, such as the location changes of people, AGV, materials, etc., and can send three-dimensional coordinates to the twin database in real time, and then transfer the data to the virtual workshop, the corresponding three-dimensional models in the virtual workshop synchronously change locations. The data in the workshop service system is mainly the order task, tooling management, material management and other data. Compared with the delivery and completion of the order task, the data reaches the actual workshop and also reaches the virtual workshop, thus realizing the virtual workshop real-time mapping with the actual shop floor.

Various types of information are generated all the time in the actual workshop. In order to realize the real-time mapping between the virtual workshop and the actual workshop, and to realize the data-driven model, it is necessary to summarize and sort all kinds of information in the actual workshop. For example, the information of the workpiece, first of all, to determine the real-time position of the workpiece, that is to obtain the three-dimensional coordinate information of the workpiece from the twin database, in order to drive the change of the workpiece position in the virtual workshop, as well as the workpiece in and out of the workshop signal, the current workpiece's position, the current process of the workpiece, and all the operations of the workpiece are shown in Table 1. In addition to the workpiece information, there are information such as machine tools, people, AGV, etc.

Table 1. Workpiece information table.

Data Name	Data Form	Return Value	Change Node	Function
Workpiece	3D coordinates	Workpiece position in real time	Position change	Drive
	0/1	Workpiece in and out of the shop signal	Whether the workpiece is in the workshop	Drive
	Text	Workpiece current station	Station change	Show
	Text	Current process of the workpiece	Process change	Show
	Text or flow chart	All processes of the workpiece	No	Show

3.2. Digital driving simulation control technology based on digital twin

AGV is an automated transportation device that can perform specified transportation tasks in a given path and scene information layout, which is of great significance for improving the production efficiency of intelligent manufacturing units and reducing production costs. Therefore, this paper proposes a digital driving simulation control technology based on digital twin. First, it is necessary to build a virtual workshop in Unity3D that is consistent with the actual shop floor, and bind the state machine and data drive conditions to the AGV model. When the condition is met, the AGV will run on the specified path. Secondly, it is also necessary to use the AGV path planning algorithm to formulate the AGV trajectory and obtain the optimal solution of the AGV trajectory. The main process is to first judge whether the task queue is empty. If it is not empty, the first task needs to be selected according to the task priority. Perform path time calculation on the task to determine whether there is overlap in the time window. If there is no overlap, the node registers the AGV and the time window. If they overlap, the time window is advanced or pushed back to find the idle time. The process is repeated until the task team is empty. Finally, the simulation runs in the virtual workshop according to the results, and the node and time information are output after the end. The realization process of the simulation control technology based on digital twin is shown in Figure 3.

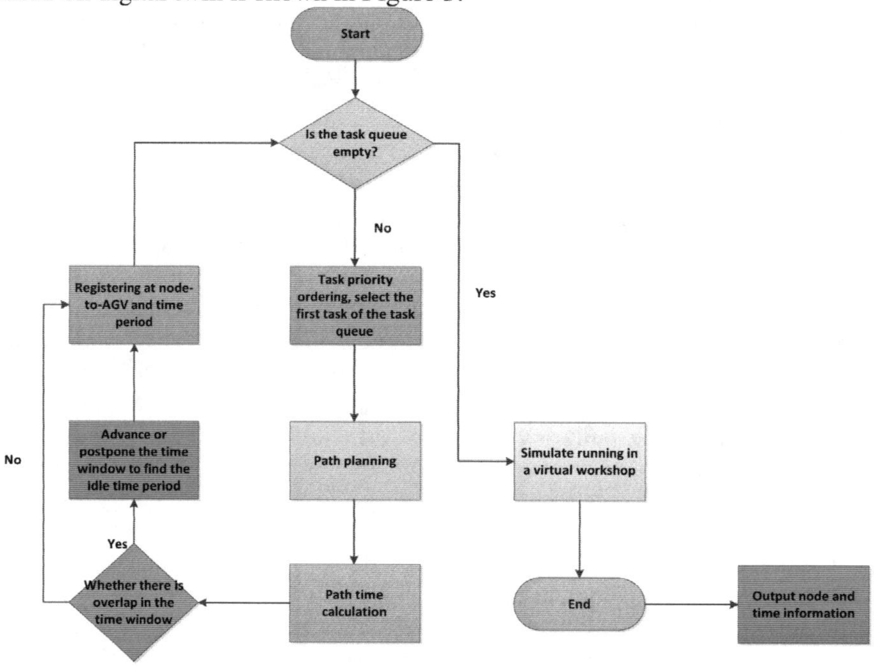

Figure 3. Implementation process of digital driving simulation control technology based on digital twin.

4. Experimental verification

This paper studies the digital twin process and simulation operation control technology for intelligent manufacturing unit. Based on this, combined with the development of intelligent manufacturing unit management system, a corresponding system is designed, and the real-time mapping technology between virtual workshop and actual workshop is solved by this system. Based on the application of digital twin simulation control technology in actual production, the virtual workshop is built to solve the real-time visual control and AGV simulation operation control of the intelligent manufacturing unit control system for the actual workshop.

In this system, the data in the actual workshop, such as AGV, staffs and workpiece changes, can be transmitted to the system through the twin database, and finally to the virtual workshop, to achieve a series of changes in the model flow, location update, etc., thereby to achieve real-time mapping purposes. As shown in Fig. 4 and Fig. 5, it is also possible to control the path simulation of the AGV, calculate the optimal path of the AGV operation and simulate in the virtual workshop to realize the simulation operation control of the AGV based on digital twin.

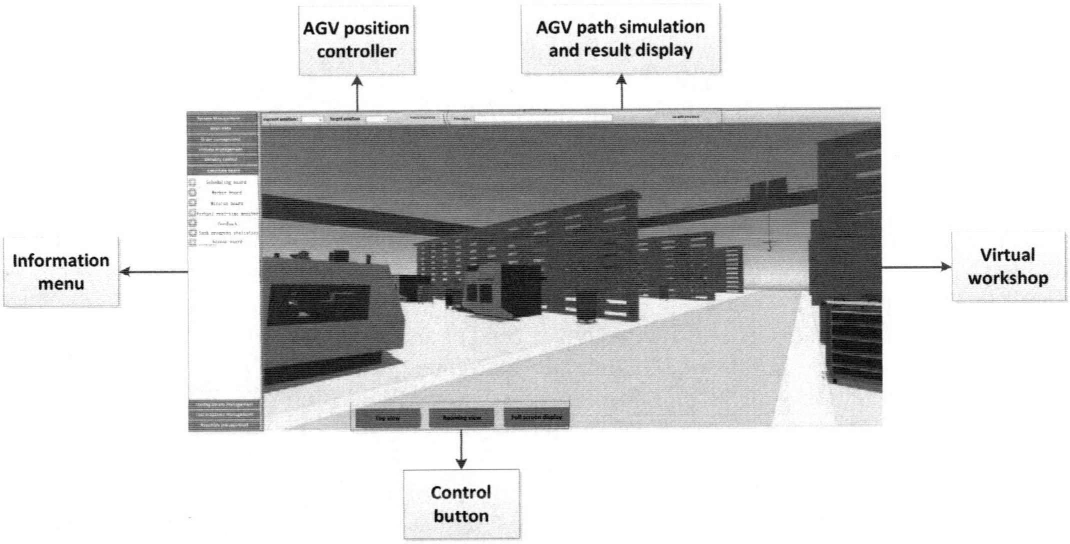

Figure 4. Schematic diagram of simulation operation control 1.

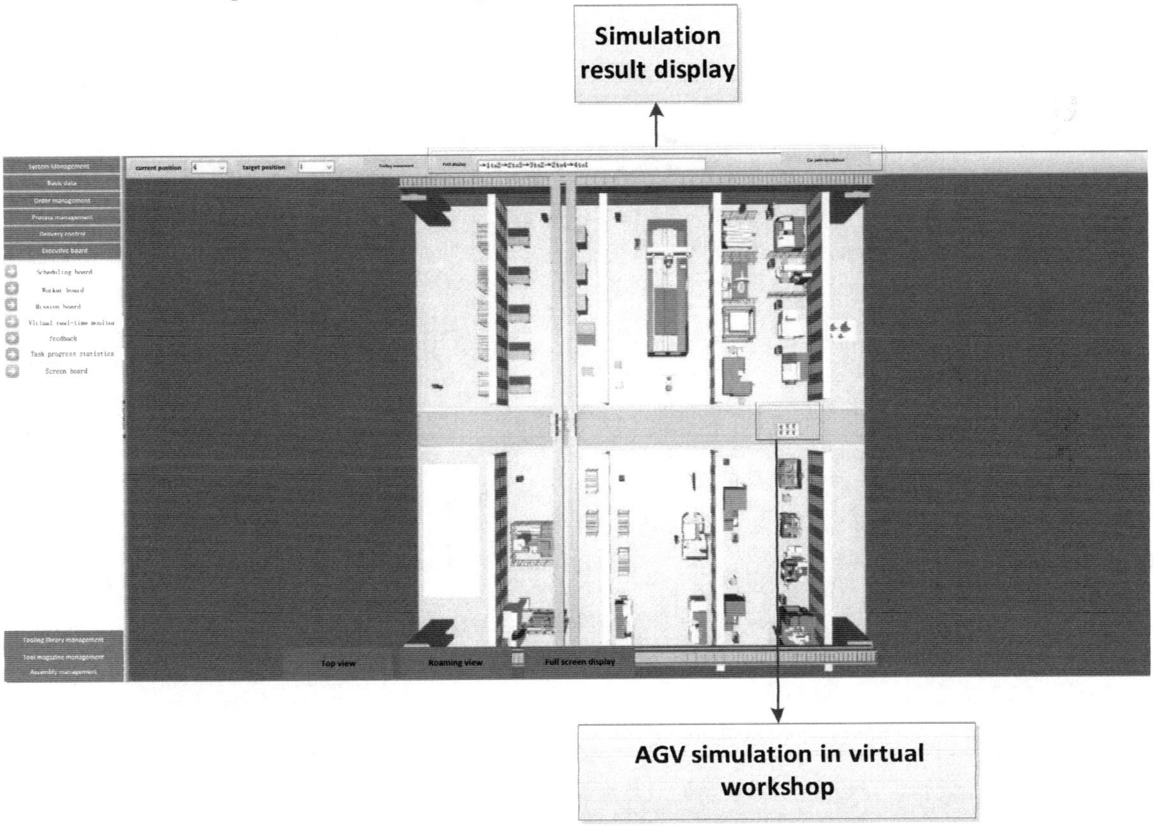

Figure 5. Schematic diagram of simulation operation control 2.

5. Conclusion

This paper studies the digital twin process and simulation operation control technology for intelligent manufacturing unit. Based on this, combined with the development of intelligent manufacturing unit control system, a corresponding system is designed, and these two technical means are applied to

IOP Conf. Series: Materials Science and Engineering **836** (2020) 012010 doi:10.1088/1757-899X/836/1/012010

establish the virtual workshop and the corresponding twin database for the purpose of solving the real-time mapping between the actual workshop and the virtual workshop and the simulation control operation based on the digital twin-based vehicle path, which improve the digital level of the intelligent manufacturing unit and the operating efficiency of the AGV.

6. References

[1] Preimesberger C. GE Unveils a Ton of new Predix Products for Industrial IoT[J]. Eweek, 2016.
[2] Zhuang C, Liu J, Xiong H, et al. Connotation, architecture and trends of product digital twin[J]. Jisuanji Jicheng Zhizao Xitong/Computer Integrated Manufacturing Systems, CIMS, 2017, 23(4):753-768.
[3] TAOFei, CHENGYing, CHENGJiangfeng, et al. Theories and technologies for cyber-physical fusion in digital twin shop-floor[J]. Computer Integrated Manufacturing Systems,2017,23(8).
[4] Tao F, Zhang M. Digital Twin Shop-Floor: A New Shop-Floor Paradigm Towards Smart Manufacturing [J]. IEEE Access, 2017, 5:20418-20427.
[5] Fourgeau E, Gomez E, Adli H, et al. System Engineering Workbench for Multi-views Systems Methodology with 3DEXPERIENCE Platform. The Aircraft RADAR Use Case[M]// Complex Systems Design & Management Asia. Springer International Publishing, 2016.
[6] Grieves M, Vickers J. Digital Twin: Mitigating Unpredictable, Undesirable Emergent Behavior in Complex Systems[M]//Transdisciplinary Perspectiveson Complex Systems. Berlin, Germany: Springer-Verlag, 2017.

AUTHOR INDEX

Institute of Physics
Dirac House, Temple Back
Bristol BS1 6BE UK

CURRAN ASSOCIATES INC.
proceedings.com

ISSN: 1757-8981
ISBN 978-1-7138-2679-8